THE CHINESE SPACE PROGRAM: A Mystery Within a Maze

Joan Johnson-Freese

AN ORBIT SERIES BOOK

KRIEGER PUBLISHING COMPANY

MALABAR, FLORIDA
1998

Original Edition 1998

Printed and Published by
KRIEGER PUBLISHING COMPANY
KRIEGER DRIVE
MALABAR, FLORIDA 32950

Library of Congress Cataloging-In-Publication Data.

Johnson-Freese, Joan.
The Chinese space program : a mystery within a maze / Joan Johnson -Freese.
p. cm. — (An orbit series book)
Includes bibliographical references and index.
ISBN 0-89464-062-3 (alk. paper)
1. Astronautics—China. 2. China—Foreign relations—1976- I. Title. II. Series.
TL789.8.C55J64 1998
387.8 0951—dc21 98-3601
CIP

10 9 8 7 6 5 4 3 2

Series editors
Edwin F. Strother, Ph.D.
Donald M. Waltz

Contents

Acknowledgments

There are several individuals and institutions that have been very supportive regarding the research that went into this project. I would like to thank the Air War College for allowing me to participate in the Regional Studies trip to China in 1997, specifically Dr. David Sorenson for his support in making that possible, and Cols. Dale Autry and Len Lyles for allowing me to join their trip. Also, I am grateful to the Air War College students on that trip who offered their support, suggestions and insight, and Admiral William Pendley USN (ret) and Lt. Col. Rob Preissinger for their pictures! I would also like to thank the USAF Institute for National Security Studies (INSS) and the Air University Foundation for their financial support in making this project possible through travel funding for a second trip to China later in 1997. From my perspective, visiting China and actually experiencing the Chinese culture, talking to the Chinese people, and conducting interviews considerably added to whatever analysis this book is able to provide. On that second trip, my son, JB, and George Moore were my constant sources of support and encouragement. They both endured my grousing and shared unexpected adventures, especially mountain climbing on horseback. George keeps me on track and endlessly reads my material, tactfully making on-point suggestions, usually convincing me that they were my ideas to begin with. I would also like to thank Mr. Chris Lanzit at Hughes Asia Pacific, Ltd. for his continued assistance, willingness to critique my writing, and invaluable insight. Dr. David Kendall and several others at the International Space University also provided comments, as did Dr. Roger Handberg at the University of Central Florida, my long-time friend, sometimes coauthor and, gratefully, never-subtle critic. I am grateful to Mr. Baosheng Chen at the China Great Wall Company in Washington, D.C., for his willingness in trying to arrange interviews for me in China, and with technical information. Chang Hui-Tzu read a draft of my manuscript and assisted considerably when I was confused about the Chinese language. Mary Roberts, my editor at Krieger Publishing was, as always, a pleasure to work with. I am very appreciative to all of these people, and many unnamed others. Clearly, however, the responsibility for the material written, interpretations, and analysis remains my own. Finally, the writing process is always long and arduous. It is often made easier with friends nearby; such was Tiger.

The Author

In 1998 Joan Johnson-Freese joined the faculty at the Asia Pacific Center for Security Studies in Honolulu, Hawaii. From 1993–1998 she was a professor of international security studies at the Air War College, Maxwell AFB, Alabama. Prior to that time she was director of the Center for Space Policy and Law at the University of Central Florida and a member of the Political Science Department faculty there. She has taught space policy at undergraduate, graduate, and Professional Military Education (PME) institutions, and has served as the cochair of the Policy and Law Department of the International Space University summer session program since 1994. She holds a Ph.D. in political science from Kent State University. Her professional experience includes: Office of Technology Assessment, U.S. Congress, Advisory Panel for U.S. Space Launch Capabilities Study 1994–1995; Guest lecturer on Space Policy, Space Tactics School, 1996; Army War College, 1994, 1995; Air Force Academy 1994; SPACECAST 2020, Advisory Board, 1993–94; and National Academy of Sciences, Space Science Board, Committee on International Programs, 1994–current. She is the author and coauthor of numerous books and journal articles. Her books include *Changing Patterns of International Cooperation in Space* (1990), *Over the Pacific: Japanese Space Policy into the 21st Century* (1993), and with Roger Handberg as coauthor: *The Prestige Trap: A Comparative Study of the U.S., European and Japanese Space Programs* (1994) and *Space, the Dormant Frontier: Changing the Space Paradigm for the 21st Century* (1997).

Chapter 1

Introduction

The Mystery

In 1972 Barbara Tuchman wrote about China that: " . . . ignorance of the language is a barrier equal to being deaf. A six-week visitor under this handicap can offer conclusions as impressions only."[1] Beyond the handicap and very real barrier imposed by language, however, are myriads of other obstacles to be encountered when researching China. The consequence of those difficulties dictates that a research piece such as this, undertaken by a space policy analyst rather than a China expert over the relatively short period of time of 18 months (according to Chinese standards and expectations), yields an analysis more limited to generalizations than usual, albeit hopefully useful generalizations.

China is truly a country of complex extremes, sometimes carefully counterbalanced, sometimes actively subverting each other. Each visitor has personal impressions of China, most of them likely accurate at least in part. The vastness of China enables nearly every hypothesis about it to be confirmed: if one goes to China looking for an open, thriving economy, evidence to that point is ample; if one goes to confirm the oppressive, closed nature of the society, evidence exists for that too. The complexity which inherently has characterized China for more centuries than most countries have known histories is now further compounded by the dynamic economic growth being experienced daily, changing the environment in which events transpire.

The difference between analyzing other countries and China can perhaps best be described by an analogy with games. Whereas chess is a favorite game of strategy in many Western countries, *Wei Qi* (pronounced "Way-Chee") is the strategic game of choice in China. In chess, each opponent has 16 pieces to maneuver toward victory; in *Wei Qi* each opponent has 256 pieces with which to strategize. Subsequently chess, considered complex to the extent that IBM awarded $100,000 to three people in 1997 as a prize for being able to program its Deep Blue computer to beat chess champion Gary Kasparov, becomes a game of almost direct attack compared to the opaque and ambiguous nature of potential moves within *Wei Qi*. Hence the degree of exponential analytic complexity faced by those exploring China today.

There are a few sagacious, but dated, unclassified publications concerning Chinese space activities. In particular, a 1992 article by John Wilson Lewis and Hua Di, based on interviews conducted in China over a 10-year period, offers astounding detail and insight into the workings of the bureaucracy and the Chinese ballistic missile program.[2] Otherwise, comprehensive written materials on the decision-making aspects of Chinese space activities, as opposed to descriptions of hardware being built or in operation, are few[3] and even many of the descriptive pieces are produced by journalists or industry groups in conjunction with escorted public relations or goodwill tours, which of course showcase and emphasize what the Chinese choose to show. There are also a limited number of articles in Western publications provided by the Chinese themselves describing milestones, hardware, and official intentions.[4] But in terms of comprehensive written analyses from analysts able to take a candid, open look at the program, there are no more of those available than there were concerning Soviet space activities in 1960, 1970, 1980, or even 1985. Indeed the parallels between China and the former Soviet Union are several.

The Chinese space program has been described as "shrouded in mystery until recently."[5] I would suggest that the shroud has been lifted only slightly, and what is being exhibited is carefully orchestrated. Looking at the Chinese space program, literally or figuratively, is much like viewing a Disney attraction where one is shown an impressive display, but also has the feeling that looking behind the curtain might reveal a reality that is very different from the public facade. After considerable time and effort on my part, perhaps my most confident conclusion concerning China today is that the real challenge from a researcher's perspective is to look behind the curtain and sort reality from illusion.

Space is plainly important for China. Beijing hosted the 1996 meeting of the International Astronautical Federation (IAF), the premier international space conference. A prior attempt to host that conference in 1989 was foiled by the events at Tiananmen Square. In 1996, the importance China attached to hosting the gathering was evidenced by the appearance of Jiang Zemin, president of the People's Republic of China, addressing the attendees. In his speech, the president said that China is open to a wide range of international cooperative efforts in space and is making space technology a high national priority in support of economic development. Wang Liheng, vice president of the China National Space Administration (CNSA), at the same meeting reinforced the theme that "China promotes international cooperation in space technology and its applications."[6] This is one theme I found the Chinese reiterating in every available forum. Along with the United States, Europe (taken together as the European Space Agency, ESA), Russia, and Japan, China is a major space-faring nation, with indigenous capabilities and offering launches on a commercial basis. China was also the third country in the world to achieve the technology necessary for recoverable satellites, after the United States and the Soviet Union, evidencing the maturity of the program in at least some areas.

All of this makes it imperative that a better understanding of how the Chinese space policy decision-making apparatus works be developed, and consequently an understanding of how to work with the Chinese is clearly needed. As John Lewis and Hua Di noted earlier: " . . . the technologies, strategies and goals relating to Beijing's missile programs must be better understood by the concerned international community in order to overcome its confrontational stance with China and to build a cooperative regime."[7] The same reasoning applies beyond the missile program to space generally. Chinese military strategist Sun-Tzu, incumbent around 512 B.C., wrote in his book *The Art of War*, "know your enemy and you will win a thousand battles." Today, the premise must be extend to "know your partner," or if not your partner, at least your counterpart, in order not to inadvertently cause unwanted confrontation. Hence my decision to undertake this project, obstacles and all.

It is not my intent to attempt to offer either the detail or the acumen of the earlier Lewis works.[8] Rather, it is intended to serve the purpose of facilely explaining, as much as is possible, a system inherently complex because of historical and culture factors, and then deliberately exacerbated in its complexity for the sake of outsiders seeking to peer in: a mystery within a maze. Further, in Chapter 6 it is my intent to extrapolate from the information provided and project where China might be going in the future and equally important, what policy actions the United States might take to avoid a confrontational stance with China while encouraging Beijing to build a more stable, cooperative regime.

The Chinese space program in 1997/98 can be said to be in some respects a microcosm of China as a whole. China is simultaneously considered a world power, and a developing country; the Chinese space program is simultaneously a world class space program by many standards, and a program dependent upon meager internal resources and foreign assistance for its future development. There is potential for great leaps forward in the future in both cases, but the schizophrenic nature of life in the near term can also be dangerous due to its instability and growing pains. Perhaps one aspect of the Chinese culture which could be positive in this whole situation is that the people are culturally more patient than, for example, Americans. Although the U.S. Congress demands China change overnight, the Chinese seem willing to proceed one small, incremental step at a time. Time may be the one way to nurse China from this schizophrenic condition to health, rather than madness.

It is my thesis that because China is at a precarious point of development, careful analysis is warranted within the U.S. policy community toward identifying ways in which the United States might urge China toward a more acceptable position on the international political spectrum. Specifically regarding space, China has traditionally emphasized the military and arms sales potential of space and space hardware. The United States ought to focus on the development of constructive US policies aimed at encouraging China to alternatively emphasize space as an element of peaceful domestic development. These policies ought then be vigorously pursued by the United States. The time is now propitious because China's current international stature comes more from its market size than its military position or ambitions, perhaps making it more receptive to outside influence than might be the case in 10 to 15 years if the status quo prevails.

There is a school of China-watchers which says that China has a recent tradition of being responsive to the moves of others rather than initiating strategy on its own. If this is correct, the United States ought to give special care not to inadvertently provide it with reasons to move in directions it would not otherwise. There is another school, however, which says that is not really the case, pointing to the 1979 attack on Vietnam, support of the Khmer Rouge in Cambodia against the Vietnamese, China's moves in the Spratly Islands, demonstrations of missile capabilities for the benefit of the Taiwanese and the like as evidence. If that perspective is correct, then there are areas of common concern between East and West which need to be stressed rather than pushing Beijing to feeling it necessary to pursue policy extremes. An approach of this type would also encourage Beijing to become more accommodating to foreigners not from a domineering, nefarious sense, but with the same kind of respect that they are now rightfully demanding.

Research Hurdles

There were significant similarities between this project and an earlier study I undertook concerning the Japanese space program,[9] particularly regarding the notable impact of culture. Oriental cultures are very different from anything either Americans or Europeans are usually familiar with, especially on matters such as decision-making processes. The differences, however, were also striking. Retrospectively, in Japan I was *just* trying to penetrate almost impregnable cultural and language barriers, and I was doing it with the full support of the Japanese government. In China the same cultural and language barriers were well in place, overlaid by very real laws protecting Chinese national security, and I had not been invited by the Chinese to undertake the project. The Japanese are constitutionally prohibited from having a military space program, making the national security aspects of Japanese space activities flow primarily from commercial economics, rather than weapons systems or strategies. This is not to underrate the importance of commercial security issues,[10] it is only to indicate a difference between the two countries. In China, civil and military space activities are basically symbiotic in many cases, separated only at the applications level. Subsequently, talking about even the most seemingly innocuous aspect of decision-making, ownership, and responsibility chains in the civil space realm can be classified in China. In other words, nothing is made easy. China is not "user friendly" when it comes to either seeking or finding information on any subject, but particularly those such as space.

Space activity, because it shares symbiotic technology in both the civil and military applications, is a tightly held field. The degree of paranoia about potentially divulging something officially considered related to national security, rendering those who might talk about it susceptible to prosecution for divulging state secrets, is significant. Subsequently, a "better safe than sorry" attitude about not talking to foreigners about space activities is always the prudent one for the Chinese. Iris Chang, writing about Chinese space pioneer Qian Xuesen (H. S. Tsien), described the sense of secrecy and paranoia that she faced during her first visit to China in 1993. "I became keenly aware of this secrecy when I was invited to a dinner in Beijing at which certain colleagues of Tsien's pleaded with me not to write anything that would offend Tsien for fear that they might be punished."[11] She mentions others who have had similar experiences as well.

As my own research progressed it became increasingly apparent why it took John Lewis and Hua Di 10 years to complete their 1992 article. There is a tacit expectation on the part of the Chinese that similar periods of time will be devoted to nurturing relationships in China on all research projects before forthright and substantive insight is volunteered, if at all. But even then, there is no guarantee, and for very self-protective, understandable reasons. On more than one occasion I was told during interviews in China that Hua Di is considered a traitor in China for his part in writing about strategic Chinese programs. Because of the degree of specificity of the information in the article, it is not unlikely that those associated with him have been censured also. Subsequently, although the Chinese have an expression, "The first time we meet we are strangers, the second time we are old friends," on a subject as sensitive as space, I would posit that may or may not be the case.

As a corollary to the classification issue, in China, the availability of documentation about space (or military) activities is tightly controlled. Whereas in the United States one's office can become overwhelmed with paper in the course of research and writing, it was my experience that receiving a piece of paper at a briefing or meeting in China without a backdoor connection (*Guan Xi*, more on that later) or significant prodding (which raised apprehensions) was rare. Comparing my experience again with that of biographer Iris Chang, she said, " . . . the information available is sparse, three years of research has pieced together an obscure journal article here, an official history there."[12] When documentation was offered, it was more likely to be at a commercial rather than a government facility, and then still, only on a comparatively frugal basis. Gifts are always given, but not documentation. Add to this a deliberate and structured system of internal information control, and one finds that people are familiar with aspects of their job on almost a "need to know" basis. This is one of many similarities with the old Soviet system.

Compartmentalization is problematic at every level in China, from operational to strategic. It is a problem not just for outsiders seeking to look in, but for the Chinese themselves, though that is only admitted privately. At the operational level, for example, it was initially speculated by the Chinese themselves that one of the recent failures of the Long March (LM) launcher was attributable to the incompatibility of a plug and socket. It seems that the two components were made in different facilities in China and when a change was made in one, compartmentalization kept the knowledge of the change from being shared with the other. Subsequently, that turned out not to have been the cause of the failure, and that it had been even considered was later denied. At the strategic level, from an analyst's perspective sorting through whether mid and senior level managers know a piece of (innocuous) information and cannot or will not tell you, or if they simply don't know, is frustrating at best.

For example, in Yanping Chen's 1993 article in *Space Policy*, she cites the Space Leading Group in the State Council as "the top group responsible for policy making and mission coordination among the central government agencies."[13] As a policy analyst the role of that group would obviously be of interest to me. However, upon asking approximately 10 Chinese officials about the role of this group, including a vice president of the China Aerospace Corporation (CASC), all either declined to answer, ignored the question, or claimed there was no such organization. That same vice president, when asked about the wiring diagram of his own organization, professed not to know who his boss worked for in terms of policy guidance and funding. Although by Western standards this seems ludicrous and one is tempted to assume that this was simply a case of stonewalling, in China his claim not to know could well be true. Other possible reasons for the apparent confusion have been suggested by other space researchers as well, including: " . . . a tendency of the Chinese government to change the names of institutes and missiles; the different Chinese, American, and Soviet designations for the same missile; [and] faculty translations that render historical descriptions vague and confusing."[14] Whatever the real problem, ambiguity is simply a hurdle with which to contend.

According to Confucian tradition, there are no "individuals" in China. Everyone is linked in an intricate web of relationships, as someone's daughter, son, mother, brother, and so on. These relationships by which all interaction is defined are called *Guan Xi* ("Gwan-Shee"). It basically says that although a wiring diagram may say D reports to C, C reports to B, and B reports to A, and A is the boss, D is A's brother-in-law and they have dinner twice a week and D is the real route to getting things done with A, not B or C. The same premise of "linked relationships" extends today to foreigners. Other than those traveling to China on a tourist visa, visitors must be "invited" to China by a Chinese sponsor, which then becomes the basis for the foreigner's identification. Who sponsors a visitor becomes the basis for judging the visitor's position, and hence the obligation of the Chinese to respond to them. Without a sponsor, chances for substantive meetings with individuals in Chinese organizations, public or private, are greatly reduced. Further, in some cases meetings between Chinese and foreigners must be coordinated by the Foreign Ministry, and often foreigners are accompanied to their meetings by Chinese Foreign Affairs officers. Once at the meeting, information is carefully released if at all, again, for fear of divulging something classified.

That being said, the research for this article was then primarily accomplished by a combination of induction and deduction. Bits and pieces of information were gathered from a variety of sources, pieced together and compared, and extrapolation drawn. In the course of two trips to China to conduct interviews, I found that in many cases I could learn as much from those who had worked with the Chinese, as from the Chinese themselves. In some cases, even American businessmen seemed reluctant to discuss their dealings with the Chinese because of, as it was stated to me, "a sensitive climate." The Chinese people I worked with were, down to the person, polite, friendly, and proud of their country. The "handlers" from the Chinese Foreign Ministry who were in charge of orchestrating and supervising my first visit as part of a larger group advised me to ask any question that I wanted, and to stop when people told me they were unable to answer. Although helpful, these handlers clearly were also responsible to report on my activities and interviews. Often, I felt that the people I was interviewing would have preferred for me to simply make something up about their space program, rather than being asked to talk about such a "difficult" subject. Indeed in one case, a person came right out and suggested that to me. Another suggested that I write my article and send him a copy after it was published and he would then tell me if it was accurate. When I asked if he would tell me how to correct my errors, he merely smiled noncommittally.

One may be tempted to dismiss my experience as attributable to my affiliation with the U.S. military as a faculty member at one of its professional military education institutions. Although that may be the case, I would encourage others to pursue similar research paths and compare the results. I would suspect, unfortunately, that both the experience and the results would be strikingly consistent. Indeed in the course of my research I spoke with space researchers in multiple fields from multiple countries and there was near unanimous agreement: at a one-on-one level, Chinese might be willing to be candid, but officially barriers are still firmly in place. Part of it, again, is cultural.

Vice Minister of the Chinese State Education Commission Wei Yu (roughly equivalent to the U.S. Department of Education) in a 1996 interview with a journalist, expressed concern about the spread of information generally. "Toward the end of our session, I was startled to hear her declaim against the growing availability of uncontrolled information—from sources like CD-ROMs, the Internet, and even CNN! 'Don't you think it's corrupting,' she asked indignantly, 'to give so much information to the population so fast? There are no secrets. We shouldn't deliver so much to the whole world. People will be confused, they will get wrong ideas.' "[15] Clearly, information about Chinese space activities has to date been scarce, and both overtly and subtly controlled.

Conclusions

I view this work as a beginning, a rudimentary guide, for other researchers, with both my encouragement and solace. Language was a barrier for me and likely will be for many others as well. Still, valid and valuable research can be carried out. From an analytic perspective, however, the key requisite for relevant and valid research seems to be viewing the issues through the appropriate cultural prism. Likely a clear view of the Chinese psyche is difficult if not impossible for Westerners, but a glimpse through the prism is certainly possible. For that reason, I suggest that anyone who undertakes further research in this area begins by reading as many books as feasible on Chinese culture.

Endnotes

1. *Notes From China*, (New York: Collier Books, 1972) 4.
2. John Wilson Lewis and Hua Di, "China's Ballistic Missile Programs, *International Security*, Fall 1992, Vol. 17, No. 2) 5–40.
3. Anne Gilks, "China's space policy: review and prospects," *Space Policy*, August 1997, 215-227; Yanping Chen, "China's Space Interests and Missile Technology Controls," *Space Power Interest*, ed. Peter Hayes (Boulder, CO: Westview Press, 1996) 71–84; Yanping Chen, "China's Space Commercialization Effort, Organization, Policy and Strategies," *Space Policy*, February 1993, 45–53; Yanping Chen, "China's Space Policy, A Historical Review," *Space Policy*, May 1991, 116–128; Gordon Pike, "Chinese launch services: A user's guide," *Space Policy*, May 1991, 103–115.
4. See, for example: Zhu Yilin and Xu Fuxiang, "Status and prospects of China's space programme," *Space Policy*, February 1997, 69–75; Liu Ji-yuan and Min Gui-rong, "The progress of astronautics in China," *Space Policy*, May 1987, 141–147; Wu Guoxiang,

"China's space communication goals," *Space Policy*, February 1988, 41–45; He Changchui, "The development of remote sensing in China," *Space Policy*, February 1989, 65–74.

5. Pierre Langereux and Christian Lardier, "Launch setbacks fail to dent China's space ambitions," *Interavia*, December 1996.
6. Peter B. DeSelding, "Chinese Set Ambitious Plans," *Space News*, 14–20 October 1996, 19.
7. 1992, p. 5, referencing John W. Lewis, Hua Di and Xue Litai, "Beijing's Defense Establishment: Solving the Arms Export Enigma," *International Security*, Spring 1991, 97–109.
8. Also including John Wilson Lewis and Xue Litai, *China Builds the Bomb*, (Stanford, CA: Stanford University Press, 1988).
9. Joan Johnson-Freese, "*Over the Pacific: Japanese Space Policy Into the 21st Century*," (Dubuque, Iowa: Kendall-Hunt, 1993).
10. John J. Fialka, in his article "While America Sleeps," *Washington Quarterly*, Winter 1997, focuses on espionage taking place within the United States beginning at the university level and expanding into the business arena, particularly by Pacific Rim countries. Of China he says, "Although the FBI makes an effort to watch foreign students and business people, China's flood has simply overwhelmed the bureau." (p. 60)
11. Iris Chang, *Thread of a Silkworm*, (New York: Basic Books, 1995) xv–xvi.
12. Chang, 209.
13. Yangpin Chen, 1993, 48.
14. Chang, 209.
15. Arthur Fisher, "A Long Haul for Chinese Science," *Popular Science*, August 1996, 42.

Chapter 2

History: Politics and Culture Intertwined

Introduction

With a history of just over 200 years and a "culture" that is usually measured in decades at most and others say has yet to develop at all, America is a country difficult for outsiders to understand. The reason for this difficulty likely stems to a significant degree from their resistance to believing that Americans are as culturally superficial and anomalous as they initially come across. To our own amusement, however, satirist Dave Barry regularly and effectively points out in his newspaper column that America is indeed a country somewhere between culturally maturing and a cultural wasteland. In fact, one need not long ponder the meaning or linkage of the home shopping channel, full combat wrestling, tent revivals, or rap music to American historical roots, but only to realize that Americans will embrace it for some amount of time, and then reject it like disco or leisure suits and move on. What the 1920's were for flappers and bathtub gin, the 1960's were for hippies and marijuana, but what either says about America or American culture has yet to be determined.

China, on the other hand, has some 5000 years of continuous history. The enormity of that fact takes a moment to absorb and to appreciate in terms of the resultant depth of its cultural heritage. That it has the world's largest population and oldest culture, and the ensuing conviction of the Chinese occupancy of the Middle Kingdom, has a profound influence on the Chinese approach to life. A country can easier change its economic structure (as China has) or its political system (which it has not) than its culture, and the depth of its culture is related to the length of its history. Hence it is important to link the seemingly unrelated concepts of culture and space policy in this book, especially to a readership likely more familiar with technology than Chinese tradition. It might be generally possible to grasp the mechanics of the Chinese space program without the benefits of historical information, but the likelihood of truly understanding the policy aspects without this contextual information is significantly less, and attempts at analysis and extrapolation become superficial at best. This is my explanation, but not an apology, for what some may consider too much historical prologue before ever getting to the subject of space specifically.

Five Thousand Years of History

One of the first Jesuit missionaries to China explained the notion of the Middle Kingdom as follows.

> One must realize that the Chinese, supposing as they do that the Earth is square, claim that China is the greatest part of it. So to describe their empire, they use the word *t'ein-hia*, "Under the Heavens." So, with this admirable system of geography, they were able to confine the rest of humanity to the four corners of their square.[1]

Being in the center of the world then inherently meant that everyone else was on the periphery, not as important, significant only in terms of their relation to China. Embedded in centuries of history and generations of thought, it is then not surprising that the Chinese have described themselves as the "first civilization on Earth," the father of the "noblest people," and "the most fully human people on earth." Again, with a longer history than most, China's assumption that it is superior to and can outwait other systems is not without foundation.

Chinese history has perhaps produced more scholarly work than any other historical field, if for no other reason than the longevity of the topic.[2] I would not presume to even attempt to summarize the works of others on that topic of lifelong endeavor. It is also recognized that condensing 5000 years of Chinese history for the purposes of this work obviously gives about the same representation of the whole as one compact disc which purports to provide all the greatest classic songs ever written: everything played at the tempo of *Flight of the Bumblebee.* It is necessary, however, to provide selected information about Chinese history relevant to contemporary internal Chinese decision making and worldview. Therefore, apologies are offered in advance for the centuries or millennia that are neglected in this overview.

A historical overview of China without a consideration of the influence of Confucianism would be akin to building a house without a foundation. Confucius, a traveling philosopher of the fifth century B.C., advised leaders on how to assure order and prosperity in their provinces, based on three core principles. Even during the time of Confucius, China was plagued by internal strife directly impacting its productivity, the welfare of the people, and subsequently the longevity of the leaders' reign. Hence offering order and prosperity to these individuals was like offering an elixir from Heaven. The three core principles were adopted from the highest levels of government down through the family units. First, Confucius preached a conservative ideology, looking not progressively forward, but seeking to recapture a mythical perfect past. Hence anything "new" becomes inherently untrustworthy and undesirable. Second, Confucianism favors strongly hierarchical political and social organization. This hierarchy creates sets of obligations according to status, with the lesser parties owing obedience and respect to those above them, and establishing a sense of obligation for those above toward those below. Those capable ought to rule. In old China, capability was in part judged by written examinations about Confucian teachings. This examination system created the basis for the bureaucracy which quickly became, and remains, extremely powerful. Finally, the core of Confucianism is a requisite understanding of "correct" conduct by all members of the population; the governing of the hierarchical relationships is by rules and standards. One's responsibility is to a limited number, but absolute within that number. It is from this core principle that the "web" notion referred to earlier is from, and the importance of the "web" then lessens the importance of individuals. It is imperative to remember that the goal of Confucius and his teachings was social harmony (stability). Clearly, according to Confucianism all are not created equal and hence democracy has no ideological roots in China. Also, there is little need for interest beyond one's web of responsibility, creating not only a sense of isolation, but also a desire for isolation.

The Chinese have isolated themselves with great deliberation. Until recently, the Chinese had been described as not really curious about anything not Chinese. The Great Wall, with antecedents dating back to the year 214, acted as a second line of defense behind the belt of deserts and high mountains that surrounds China. The wall itself can be said to be made from the blood and sweat of the Chinese people not just symbolically because of the vast numbers who worked on it, but in that innumerable of those who died of exhaustion in the construction were simply thrown into the mortar to strengthen the wall against outsiders.[3]

Within this self-imposed isolation, the aspects of Chinese history relevant here are those which pertains to internal-external relations. "Recent" suspicion of foreigners by the Han Chinese, which at 91.9% makes up the majority of the Chinese population,[4] dates back to the Mongol invasions in the 1200's. The legacy of Mongol rule, which took the Chinese by surprise, took multiple forms, including a more strongly centralized and autocratic form of government, restriction of individuality and innovation, and a more culturally introverted Chinese nature.

China returned to Chinese rule in 1368, with the rise of the Ming dynasty. Until very recently, Chinese expeditions into the world of the barbarians beyond were few and brief. In the 15th century a series of expeditions sailed to canvass foreign shores. Beginning in 1405, these were known as the cruises of the Ming admirals.[5] They ceased only a scant 28 years later with an abrupt return to isolationism. Indeed the Chinese burned their boats, said to have determined that there was nothing beyond China of interest. The price for their isolationism turned out to be high, as it was at about this same time that the Europeans began their exploration and expansion.

In 1644, the Manchus were poised at the Chinese frontier ready to attack. Their campaigns culminated with the rise of the Manchu Qing (or Ch'ing) dynasties, which meant that the Chinese were again being ruled by outsiders. The Manchus introduced few major economic or social changes into Chinese life, indeed to some extent there were deliberate attempts to keep the cultures separate. Intermarriage between Manchus and Chinese was officially forbidden until 1902. It was, however, the Manchus who instituted the shaving of men's heads and the wearing of queues (pigtails), much to the distaste of the Chinese people. The Manchus also tried to ban the Chinese practice of foot binding among women. So deeply embedded had the practice become (since around the 900's) that they were largely unsuccessful.

It was around 1700 that opium was introduced into the Chinese culture. Opium had been known to the Chinese several centuries earlier, primarily as a medicine to be taken internally. Until about 1523, the Chinese imported opium from India for these purposes. Coastal pirates, however, made importation so difficult that the Chinese began growing their own. By around 1700, people began the practice of smoking opium for its dilatory effects, leading to its being banned by imperial decree in 1729. Opium smoking continued, however, which led to a ban on importation in 1800. From the foreign perspective, the timing of that prohibition was displeasing. While the rest of the world was dazzled by the tea, spices, silks, and other goods to be had from China, the only thing of interest the rest of the world could offer China in return was opium. By 1800 drinking tea had become not only a British obsession, but the import duty on Chinese tea provided the British government with almost one-tenth of its total revenue.[6]

In 1839 foreign merchants, primarily British, bought particularly large amounts of opium to sell in China in response to a rumor that the importation ban was to be lifted. Quite the opposite was the case, however, and China clamped down on opium importation, including destroying some of that belonging to the importers. In the name of free trade, the British sent their fleet, and hostilities commenced. Eventually Chinese leaders, feeling insecure about their own military capabilities considering the clear superiority of Western weapons, and not wanting to risk losing face to the foreigners in battle for fear of the internal backlash which could have resulted, negotiated with the foreign powers. The Treaty of Nanjing (1842) resulted in increased accommodation of foreigners in China, including the ceding of Hong Kong Island to Britain, and payment of a $21 million indemnity.

From that time, the conditions were fixed for the thesis of foreign vileness and corruption to be proven to the Chinese. All the predictable vices associated with a tenant population blossomed. China imploded as a quasi-colonial territory of multiple countries. What is most remarkable about this period is the degree of ignorance the Chinese still suffered about the West. Although they had experienced contact with Western traders and missionaries in China for several hundred years, they had learned little about Western ways. Chinese attention was still turned inward. To avoid the appearance of weakness toward the Western lodgers, which the Chinese government again feared would lead to internal revolt, the government would take a hard line with the Western powers only to be forced to back down later. This led to increased concessions, further worsening the situation. A vicious circle was spinning out of control.

The Boxer Rebellion (so-called because of the supporters' devotion to Chinese "shadow boxing," known today as Kung Fu) in the late 19th century illustrates the degree to which hostilities against foreign incursions into China had escalated. At first, these rebellious members of "secret societies" were only surreptitiously supported by the Chinese government, including the Dowager Empress Ci Xi, in being allowed to bully and later execute missionaries, Christian converts, and foreigners in general. Railways and telegraph lines were destroyed as symbols of foreign encroachment. The culmination of the Boxer activities was holding the foreign legation quarters in Beijing under attack and siege for 55 days. By that time, the Boxers had been openly put under the control of the Qing court. Rescue troops eventually arrived and the legations were freed. The resultant damage to not only Chinese foreign relations, but also internal Chinese politics, proved to be irreparable.

Although the Qing court was restored, anti-Manchu sentiment had never totally been dispelled. Han dissatisfaction with the government was widespread. Indeed Sun Yat-Sen was one of the few Chinese leaders ever to study abroad[7], in Europe, after fleeing China in 1895 having been part of an anti-Manchu movement. Sun returned when, in 1911, the Qing dynasty virtually collapsed under the weight of years of turmoil and the Republic of China was founded.

A country, however, is not made by proclamation alone. For many years after the republic was proclaimed, a succession of warlords raped the land and people of China. Struggle for a new government system was ensuing at the same time as communism was gaining ground in other countries, such as Russia. Some Chinese leaders, like Sun Yat-Sen, viewed incorporation of the communists into the new republic as the right course. Others, like Chiang Kai-Sheck, were adamantly anticommunist. The civil war which ensued was further complicated by the Japanese incursion and occupation of the late 1930's. It was not until 1949 that the People's Republic of China was created, with communist Mao Zedong as its leader. Chiang Kai-Sheck, probably saved from execution by Zhou Enlai, withdrew to Taiwan with his forces, still claiming legitimacy. China, however, was back in Chinese hands.

Once the communists were in power, China turned to its brethren Russian communists for assistance. Mao, however, was contemptuous of Russian communism, much in line with the typical Chinese low opinion of most things not Chinese. Few Chinese leaders ever felt compelled to study abroad or learn the customs of others. Mao Zedong had never left China prior to going to Moscow just before the communists took power. He had not studied abroad, nor did he speak any foreign languages. He had seen no need to learn about other countries. Ties with the Soviet Union were tolerated and even nurtured, however, in return for support and assistance on such projects as obtaining a nuclear capability. It was during this period that the Chinese commenced their space program as well. In the meantime also, Mao pressed China into industrialization in the late 1950's with the Great Leap Forward, which resulted in perhaps the deadliest famine in human history. China was communist and China was in shambles.

In 1960, an irreparable doctrinal split occurred between Moscow and Beijing. China was once again on her own, and turned inward more fervently than ever before. Part of the Sino-Soviet split was doctrinal, another aspect more reflected Chinese intraparty problems. Mao considered himself China's great leader and did not accept challenges. Besides splitting with Moscow, another of his efforts to consolidate power was the so-called Cultural Revolution extending from 1966 to 1976. Although at the time the intent was said to be the purging of capitalist influences, equal or more effort was placed on purging the Chinese government of Mao's enemies. Mao, by his own design, became more than a leader of China, but a symbol of China itself. Following in the wake of other cult of personality phenomena, Mao was to China as Joseph Stalin was to Russia, "omniscient, all knowing, capable of thinking for others and accomplishing any task."[8] Actions to consolidate his own power within the Communist Party in China, and in his mind the world, resulted in big-banner symbolism and calls for reeducation that sent frenzied students into the streets with their Little Red Books of Maoisms, eager to dance, sing songs about the East being Red, and abuse anything even remotely counter-revolutionary. Retrospectively, the period has been referred to as 10 wasted years even by many Chinese. It was a period of intellectual intolerance and destruction which pushed China even further behind the developed countries. Perhaps the strongest indicator of the degree of backwardness in China exacerbated by the Cultural Revolution is that there was no compulsory education in China until 1986.

The abuses of the Cultural Revolution, particularly those of the Red Guard, eventually got to be too much even for the communist hard-liners. Inroads against Mao began to be made. In 1973 Vice-Premier Zhou Enlai, beloved by the Chinese people and "kindly mother" to Mao's "stern father" to China who had "handled" Mao since their revolutionary days together, managed to restore Deng Xiaoping to the Politburo. Deng had been expelled during the

Cultural Revolution as a counterrevolutionist bent on restoring capitalism to China. The return of Deng put him clearly in the line of succession and hence in contention with those in the "Gang of Four" (including Mao's wife, whom Mao was increasingly distancing himself from) who had plans of their own.

In 1975 at the Fourth National People's Congress, which Mao boycotted in objection to the way Chinese policy seemed to be straying, a cancer-stricken Zhou Enlai rallied himself from his deathbed to announce an ambitious program called Four Modernizations as China's goals for the next 25 years. The Four Modernizations focused on agriculture, industry, defense, and science and technology to transform China into a "powerful, modern socialist country" by the end of the century. Mao saw these goals as more nationalist than communist.[9] Among the goals, military modernization was assigned the lowest priority. It was economic reform which was to take the lead. That task was to be left to Vice-Premier Deng Xiaoping, the unsung draftsman of the Four Modernizations plan, whom Zhou had placed in the position of running the day-to-day business of China because of the infirmities of both him and Mao.

Zhou Enlai died in January 1976. Mao died 9 months later on September 9, 1976. Throughout Chinese history, the deaths of leaders have provided windows for change in Chinese politics, hence these periods are carefully handled and orchestrated. When Zhou died, crowds poured into the streets and the emotion displayed has been described as similar to that evidenced in the United States after John Kennedy died. Yet Zhou laid in state for only 2 days and then his body was cremated. Mao was not at the funeral, nor was Deng. Officially, it was ominously unceremonious.

With Mao alive there was still no vacant seat of power, but neither were there favored champions among the contenders. Zhou was dead and Mao had been incapacitated during the summer, having suffered a stroke. That left the Gang of Four, Vice-Premier and Minister of Public Security, and Mao's crony, Hua Guofeng, and Deng to maneuver for position, as it was clear that someone's time would be coming soon. Deng had been expelled from power again in April 1976, and was residing in Canton. Byzantine political strife was ripping the country apart, the economy was in shambles, and China was again on the brink of imploding.

When Mao died in September 1976, the scene was very different than it had been with Zhou. His funeral was accompanied by great ceremony and plans laid immediately for his mausoleum in the square beneath the Gate of Heavenly Peace in Tiananmen Square. Weeks of mourning were arranged and officially sanctioned, time politicians could use to consolidate power. Deng quietly moved back to Beijing. Leaders of the political and military bureaucracies moved swiftly to eliminate supporters of the Gang of Four, whose imposing personal lifestyles and ambitions transposed against their austere and choking political policies for the masses resulted in widespread animosity toward them. Indeed the fall of the Gang was engineered by Hua Guofeng and Minister of Defense Ye Jianying.[10] Although Hua remained in contention in the power struggle for some time, he was unable to rally enough support to endure. Deng emerged the victor.

Deng Xiaoping was the general architect of the economic reconstruction China is undergoing today. The plan he announced in December 1978 had three basic thrusts: ending collective farming; introducing price and profit; and seeking an infusion of foreign capital. There was, however, no general blueprint for modernization, rather it was a careful dance of incremental steps designed to push forward yet allow a pulling back if necessary to maintain stability. Hence he is credited with the "opening" of China. He has been compared to Mikhail Gorbachev in that respect. It must be remembered though that Deng, unlike Gorbachev, was acutely cognizant of the need to maintain tight control of the population, so as not to meet the same political fate as Gorbachev. The extent to which he was willing to go to maintain control was demonstrated in the bloody 1989 crackdown on the student demonstrators in the Tiananmen Square.

This is an encapsulation of the history upon which modern China is built or, more accurately, is being built. The lessons of the past are not lost on either Chinese leaders or the population. Analysts would be remiss to neglect history's importance, or that of the pervasive Chinese culture with which it shares a symbiotic relationship.

Breaking the Culture Code

The influence of history and culture on Chinese decision making, especially compared to the United States, is critical because it makes the sometimes extreme and sometimes subtle differences between the two systems understandable. While taking an acknowledgedly abbreviated and perhaps callused view of American politics, in the United States policy making can be best understood as a function of the ultimate goal of all elected officials: reelection. Within that overarching and penumbral (but realistic) view, bureaucratic politics and other systemic influences then come into play. Indeed that space policy in the United States has particularly fallen victim to that less than maximally efficient and non-achievement-oriented policy model, because the space constituency is too small to effectively demand policy accountability at a national level, was the theme of my previous book.[11] Space policy in the United States is primarily a subset of foreign policy, in the case of civil space activities, and defense policy concerning the military sector.

Politically, the Chinese system is both simpler and more complex than that of the United States, illustrating another of its contradicting extremes. The complexity derives from the power of bureaucratic politics, as a legacy of

Confucianism, which is no less prevalent today in China than in the days of the Ming emperors. As in Japan, Chinese bureaucratic politics make bureaucratic politics in the United States seem like an Olympic sport being engaged in by mere amateurs. In long-term strategic planning and posturing, Chinese politics are akin to the *Wei Qi* game reference prior. From an overarching perspective, however, Chinese politics today are strongly analogous to politics in the United States: motivated by the politicians' goal of staying in power (re-election with a Chinese twist) which they see as achievable via one, clearly set, often reiterated, simply defined path: balanced economic development. Theoretical tools within which to frame this analysis were hence sought accordingly.

In the United States, political science literature of the 1960's was rich in the work of theorists. The startling changes in the world order rendered theories like balance of power inappropriate, and trying to use them provided more anomalies that answers. Many if not most of the theories developed during that time have long since faded into obscurity after varying degrees of critique, rethought, or utilization. Indeed outside of university graduate programs in political science, most have been discarded as "political science mumbo-jumbo" by individuals or institutions more comfortable with situational analysis.

The emergence of multiple new states after World War II particularly created the need for explaining and dealing with the crowd of countries which were both unwilling and unable to play by the rules of the more developed international community. Aspects of those theories, often grouped as "developmental," were used to look at areas like Latin America and Africa. But Latin America and Africa have never been high priorities in terms of U.S. national security interests. Therefore, utilization of the development theories was even more limited than some others, e.g., reduced to almost a "fad" phase in the 1960's and early 1970's, and then abandoned in favor of more pressing national security interests like arms control, terrorism, and peacekeeping. With the breakup of the Soviet Union and the emergence of China from its self-imposed cocoon, however, these development theories have become tacitly relevant again. Both China and Russia are struggling with massive development issues, and both are priority security concerns for the United States. One development theory (actually it was called a "provisional paradigm" by its author) from the 1960's which is particularly relevant to contemporary analysis of China was put forth by Fred W. Riggs[12] and is depicted in Figure 2.1.

Riggs here basically posits that *e* equals the "degree of equality of participation of members of a population in the making of governmental decisions and in the sharing of the benefits generated by governmental activities" and that *C* equals "the capacity of a government to solve the problems confronting it."[13] *R* and *L* represent the "rightest" and "leftist" forces which struggle for

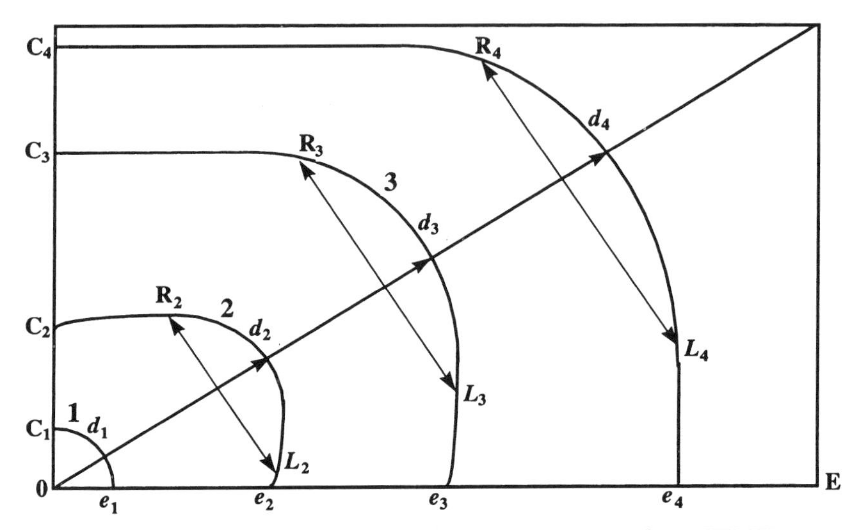

Figure 2.1 Riggs's provisional development paradigm. *Reprinted with the permission of The Free Press, a Division of Simon & Schuster from* Contemporary Political Analysis, *edited by James C. Charlesworth. Copyright © 1967 by The Free Press.*

emphasis on more capacity or more equality. With the development *d* arrows progressing from 1 through 4, the simplified point is that progress only occurs when a balance is struck between *C* and *e*. Moving outside a corridor established around the center line results in undesirable results if economic development is the desired end-state. For example, shifting too far on the side of capacity yields an authoritarian state which can stifle both the motivation of individuals to work and the ability of the system to advance. Too little capacity, however, can result in anarchy. Stability requisite for government to carry out its functions while returning tangible benefits to the people is the key to development. The applicability of these premises to the situation in China today is robust. The Chinese government must maintain its ability to deal with problems while providing returns (currently economic more than political, vice the opposite model in Russia) to the people, for development to progress, and without that development the stability of the government becomes tenuous. Stability and economic development therefore become entangled into a synergetic relationship, the hybrid combination of which is the number one priority for all decision making within China.

Within that context, external decision making becomes relatively easy to predict. To utilize another model originated within the 1960's to further focus the discussion, Graham Allison's rational decision-making model works well. In his seminal work "Conceptual Models and the Cuban Missile Crisis,"[14] Allison describes three models for consideration. Briefly, the rational model says that nations pursue courses of actions, means, to achieve identified goals, ends. The second model, the organizational process model, sees decisions less as deliberate choices, and more as routine outputs of organizations. The third model is the bureaucratic model. In it, power politics between semifeudal kingdoms and those within them become key to decisions made.

In the United States, trying to apply the rational decision-making model is difficult first because the utility of the model is null if there is no identified goal. For example, civilian space policy in the United States is burdened with multiple number one priorities or, if lumped into an ambiguous goal statement such as "maintain the U.S. position as a leader in space technology and exploration" it becomes a motherhood statement rather than anything of any meaning. At a higher level, goals become even more muddled. Are supporting and encouraging democracy and human rights goals of the United States? According to the 1997 National Security Strategy they are.[15] It certainly seems to be where U.S. rhetoric and policy toward China are concerned, but not necessarily toward other countries, like Saudi Arabia. At a lower level, the goal becomes complicated by the system itself, where all secondary goals must be balanced with the funders' overriding goal of reelection. For example, if the National Aeronautics and Space Administration (NASA) decides to move Shuttle payload processing jobs from California to Florida to save money, one of its stated and mandated-by-Congress goals, the goal can become subverted because of the reelection concerns of those same members of Congress, e.g., saving jobs in one district as opposed to moving them to another to increase efficiency. At the strategic level, China has a stated goal which it pursues by matching actions to policy, whereas the United States has multiple goals and programs which rarely match policy. Below the strategic level, however, China's system of governance is even more complicated than that of the United States simply by virtue of its decentralization, sometimes burdensome historical influences, extenuating social-economic circumstances, and most of all, its sheer size.

At the strategic level, China can be said to be pursuing with a vengeance the same development and modernization campaign which has been successful in Japan to the extent that Japan is now often referred to as Japan, Inc.. The goal is known, and rational decisions will be made within that context. By definition, a dissident is a person considered to be traitorous, or a heretic to "the system." Therefore, will the Chinese government at a time when stabilization is the number one goal disengage from those policies designed to minimize dissidents? No. Will China allow a continuation of the increasingly liberal policies and laws being passed in Hong Kong after the reversion takes place? No. Will China try to minimize the pain inflicted in Hong Kong to businesses who bring both jobs and capital to China? Yes. The predictive powers of Riggs's provisional paradigm in this environment seem amazingly accurate, far more so than were one to ask "Will the United States continue to offer China Most Favored Nation (MFN) trade status to open the lucrative Chinese market to U.S. goods?" Well, maybe. That depends on which of the multiple number one priorities professed by the United States emerges victorious on a particular day.

An assumption of this book is that Riggs's provisional paradigm and Allison's rational decision-making model are functional for analytic purposes concerning the Chinese space program. Space policy becomes a subset of more general Chinese policy within these parameters. From there, sorting through the bureaucratic politics to decipher specific decisions again becomes akin to a game of *Wei Qi* being analyzed by mediocre chess players.

Here I would also posit that the primary difference between China and Japan, both systems whose decision-making structures are laden by cultural and bureaucratic considerations beyond that of the United States, is that China, unlike Japan, does not have a legal structure to facilitate and standardize private interactions, particularly of an international commercial nature. Contracts hold as long as they benefit China or China can be persuaded to adhere to them, through carrot and stick tactics. Anyone who has ever been in traffic in any Chinese city can tell you that the Chinese regard lane lines on the road as merely advisory. If laws can be considered as analogous for regulating interaction between China and international entities, there are not even any lane markers. Further, I would suggest that although China purports to be anxious to join the Family of Nations and learn the rules of such, not having to play by any rules is currently to their benefit and hence they will procrastinate in changing that system. Several Chinese responded to my queries about building a legal system with references to observing international law, but as any who have studied international law will attest, its greatest weaknesses are voluntary jurisdiction and no enforcement power. Domestically, selective imposition of unwritten rules means that the Chinese control with whom and how they operate. If one looks at Chinese history, there is not only precedent for such, but reason for the Chinese to believe that they can outlast the rest of the world in terms of which culture will change.

Modern Challenges

Throughout this long history, two inherent weaknesses in the Chinese polity have been not only consistent, but show little signs of decline. First, although Chinese rulers have become adept and even masterful at managing the Chinese bureaucracy, they have been unable to keep power struggles from disrupting the entire system. Kenneth Lieberthal in his book *Governing China*,[16] one of the most succinct and practical books about modern China, concurrently describes that situation as well as the state of China's legal system.

> Despite myriad laws and regulations . . . the Chinese system has remained at its apex—among the top two to three dozen officials in the land—essentially lawless and unregulated. Each leader's activities at this level are restricted primarily by the attitudes and resources of the group's other members. But the inevitable conflicts and resulting instability at the apex reverberate throughout the system, disrupting the entire polity.[17]

I have already posited the state of China's legal system as being a primary differentiation between China and Japan. That premise I would suggest ought to be extended further as a critical differentiate between China and most countries it would like to deal with commercially, and a critical policy consideration generally.

So at one level, China is very organized, indeed it is micromanaged. At another, however, it is an anarchy. In international relations, this anarchical state has proven advantageous in some instances. Internally, this precarious balance within the system creates a sense of tenuousness about which leaders must constantly be concerned. In systems theory terms, where political order can range on a spectrum from stable to unstable, China is likely best characterized as "static," like a ball on a table, where even a subtle shift can propel it into instability.[18]

This leads to the second consistent weakness, that the Chinese people have no experience with political participation. Although the United States would like to see China become more democratic, the premises upon which a democracy are built are simply not evident nor even latent.

> The people of China have not . . . experienced meaningful, sustained political participation based on stable, autonomous institutions. They have often been mobilized by political authorities to exert themselves for state-directed means and, in the process, to engage in ritualistic rites of loyalty to state leaders. But truly autonomous political efforts directed toward shaping the policies and/or selecting the personnel of the state have been confined overwhelmingly to illegal underground organizations such as secret societies or to sporadic outbursts of popular protest through demonstrations and street violence. Such protests have occurred repeatedly and make China's leaders feel that radical instability—chaos—always lurks just beneath the surface calm of society.[19]

Often, the outbursts coincide with the masses sensing governmental paralysis due to conflict among the leaders (1989 with Tiananmen Square) or prodded by leaders to serve their own interests (1978, 1979). In any event, stability has long been perhaps China's most valued, and precarious, ideal.

With Deng's death, for the first time in a very long time, China is without an emperor.[20] He was the last of the strong individuals capable of exerting decisive authority over the populace. Whether admired for his economic policies or chillingly regarded for his willingness to turn Chinese troops on the

Chinese people in Tiananmen, Deng's legitimacy as a leader was respected by a population some say badly wants and needs an emperor-like figure. This view reflects back to China as a peasant society, saying that for all their new emergent modern ways, the majority of the Chinese are still peasants at heart and want someone to tell them what to do. That is why the death of leaders is such a tenuous time in China; it is an opportunity for new leaders to emerge and rally supporters during a period of vulnerability. Deng's funeral and mourning period were very low key, with sanctioned public lamentation kept at a minimum. Analysts inside and outside of Beijing agree that the new Chinese president, Jiang Zemin, is not an emperor. This is a frightening situation for both the Chinese leadership in power and the population.[21]

There is no question that the Communist Party is alive and well in Beijing. New President Jiang Zemin came to power from his position as head of the Communist Party. Indeed all those with any claim to Deng's throne are powerful party members.[22] Within the military the General Political Department (GPD), the arm of the party, is clearly the most powerful. That department is in charge of officer appointment and removal, which basically shapes the military of the future. Among the key policy players in Beijing—the party, the government bureaucracy and leadership, the military, and the new private sector—difficulties with foreign economic ventures are most often traceable back to the party, as the ultimate decision-making authority. But the Communist Party is basically under siege against a world in which communism has not only fallen out of vogue, but in some countries has been run out of town on a rail as the malefactor of all evils. Its existence, party members know, hinges on continued economic growth to keep the price of its expulsion too high.[23]

Where U.S. analysts perhaps first begin to bifurcate about China policy is on the issue of the strength of underlying support for communism from the Chinese people. While analysts are united in agreement that the economic potential of China is staggering, they are fractured on how well that transfers in a communist military threat. One group feels that few people among the Chinese population believe in or would be willing to fight for communism any longer. Unquestionably though, membership in the Communist Party is still the best way to progress on the ladder of success. Those who have the ear of the U.S. press and Congress, however, seem to feel that while the number of communists who look like they are right out of People's Liberation Army (PLA) central casting are decreasing in number, the even more dangerous ones are the young economists whose rhetoric was described by one U.S. general officer as "communism with a doily underneath." This is a portrayal of the Chinese as anchored to a bedrock of uncompromising communist beliefs and economic development as a tool for the country's advancement. I would suggest that perhaps the rhetoric that is being given by the Chinese is: first, responsive to U.S. denunciations; second, a pragmatic approach to individual career advancement; and third, more indicative of the resurgent nationalism, not communism, which is replacing communism as the dominant ideology in China. Nationalism can either be a constructive, building force, or a volatile, dangerous force, depending on how it is handled internally and externally.

Considering some recent U.S. anti-Chinese rhetoric, the Chinese may well have reason to wonder whether the United States is hostile or friendly to the Chinese. The Chinese questioned, for example, what the 1996 U.S. National Security Strategy of Engagement and Enlargement[24] meant to China, and rightly so as "engagement" is clearly not a policy, but at best an attitude. It does not address what is to be done about issues like Taiwan or the World Trade Organization (WTO). The adage "containment now" is also gaining strength among some Americans, using analogies to the rise of the Soviet Union post–World War II. Many in China see these strategies as linked to Japanese militarists. Indeed the signing of the U.S.-Japan Security Treaty in April 1996 went virtually unnoticed in the United States, but not in China.[25] Further, just like death and taxes, the annual debate in the U.S. Congress over China's MFN trade status has become a sure thing.[26] Recently, heated MFN rhetoric has had to step aside to make way for Congressional hearings about alleged illegal Chinese campaign contributions. The point is, from Beijing's perspective China has certainly garnered considerable attention in the United States recently, and none of it favorable or even neutral.

Regarding communism and career advancement, curiously, Confucianism and communism have become linked. A modern day legacy of the Confucian principles emphasizing the web of interrelated relationships is still at work in China, incorporating Communist Party membership; the *Guan Xi* discussed earlier. *Who* you know is still equal to or more important than *what* you know. While the same can also be true in the United States, and even more so in Japan, the degree of relevance is even more significant in China. Indeed the importance of power and connections, mixed with traditional Chinese cultural emphasis on family, has created a class of individuals known as the Crown Princes (*taizidang*), the Princelings, or the Cadre Kids. They are the children of the party officials, who through nepotism and blatant corruption act as leeches on society. If 10 licenses for private sector businesses become available, most or all of them may go to Cadre Kids through no competitive process. Air travel in China has been said to be more dangerous than it need be because Cadre Kids are given training slots for high-paid pilot positions not because of skill or intelligence, but because of nepotism, although in that particular field things are said to be changing for the better.

High-level corruption within China has reached a level of near intolerance by the masses. Indeed a Ministry of Supervision has been created at the national level, with extralegal authority (whatever than means in China right now) to fight corruption as an exigency issue. But since the corruption permeates all levels of the government, and the party is the strongest arm of the government, fighting corruption is like asking an animal to chew its foot off to escape a trap. This by no means implies that the plague of corruption has not permeated other societies as well. Indeed headlines in the United States and other countries are increasingly filled with cries of corruption to the alarming extent that it seems to no longer be scandal, but commonplace.[27] But "corruption" of a lesser sense, although corruption nonetheless from a Western perspective, is a way of life in China and not corruption to the Chinese, but part of their culture. As part of the culture, there is little sign that change is imminent, or even desired. Changing a process is always easier than changing an attitude, which is what is at work here.

This low-level corruption by Western standards, or standard operating procedures in the Chinese culture, references Chinese practices with regard to doing business which currently favor the Chinese. If there is no price tag on a teacup, a rug, or a rocket, then the price can be negotiated based on factors unknown to the buyer. Further, keeping the buyer as uninformed as possible is to the seller's benefit. That is the way things work in China now, which creates some serious dilemmas for Westerners trying to work with the Chinese.

There are basically two types of transactions that are critical to China today: commercial partnerships and cooperative partnerships, the latter primarily government-government. In each there are at least two parties, but motivations and goals differ. In the business relationship there is a buyer and seller; the buyer offers the market, the seller the goods. China currently has a strong hand in these relationships. The size of the market keeps the West attentive, if not beholden.[28] Foreign companies are falling all over themselves to get to the 3,300,000 individuals on the waiting list in China as of 1995 for cellular phone subscriptions, not to mention those waiting for paging service (11,210,000 in 1994) and telephone lines (1,620,400 in 1992).[29] American, European, and Japanese businesses cannot afford to wait or challenge the Chinese way of doing business; they must simply accommodate the best they can. What that most often means is that they hire a Chinese person, hopefully one with good *Guan Xi*, as their representative in China. This is their fastest route into the market, though it clearly perpetuates the Chinese methods of backdoor, *Guan Xi* laden business transactions. As a Chinese representative of one of the large American aerospace firms told me, "the West has to come to China, not the other way around."

In cooperative partnerships, however, which are far more closely akin to "Family of Nations" relationships in which China purports to want to be included, motivations and goals differ. In a government-government space partnership, for example, each party typically comes to the table with something to offer, most likely technology or money. In China's case, it has limited offerings of either, but seems to want both, preferably via "cooperation." In these cases, China is likely to claim developing country status, legitimately, where leveling the playing field for them is required. However, the idea that sometimes China wants to be treated as a world class power and other times wants special consideration can be difficult at best for foreigners, or more likely simply seen as hypocritical.

In the case of commercial partnerships, the lack of a mature legal system is often cited by the Chinese to explain their inability to accommodate normal business practices. In the case of cooperative partnerships, lack of resources is cited as requiring China to participate in international ventures as a developing country. While each may be legitimate, support for and understanding of their positions will be dependent on firm evidence that serious efforts are being made to correct these situations, particularly regarding the former. A Chinese official recently indicated that they are doing just that. " . . . At the same time, the industry is working to improve its legal and contractual framework by joining three international space and other space agreements."[30] Although clearly a first step, the measure cited again concerns (unenforceable) international law. Moving beyond rhetoric and into domestic regulation is, however, imperative. Lack of a coherent legal system is indicative of China's authoritarian status.

China today is an authoritarian state cleverly disguised as a communist country. If one wants to see true capitalism in action, go to China. Yet when President Clinton said that China was not communist in March 1997, critics severely censured him. China is guilty of human rights violations, China still has state-owned industries, and China is dominated by one party, called Communists, so they must be Cold War Communists of the "communism will take over the world" variety. If it quacks like a duck, it must be a duck. But the ensuing proclivity of the United States to view China as black or white leads to all-or-nothing policy choices which are in no one's best interests, perhaps least of all those of the United States. Further, it leads to the interpretation of events to fit into a preconceived notion. Among the admittedly limited sample of non-Chinese I spoke with in China, government officials, business people, and even tourists, all agreed that a significant portion of what appears about China in U.S. newspapers and through broadcast news outlets, some more than others is, to use a word commonly heard, worthless. More likely, it is just heavily biased toward one Chinese extreme or another. This returns to the diverse nature of China discussed prior; if one analyzes China toward proving a specific hypothesis, evidence can always be found among the expansive evidence available. What is truly frightening is the rapidity and almost zealousness with which the half-true hypotheses are then free-fall extrapolated within the United States to manufacture China into the "next big threat," an essential component for any country with a large military to justify, especially compared to the almost euphoric spirit of cooperation with which the United States is embracing the Russian government. The danger is that by taking actions based on inutile assumptions, U.S. policy-makers can create a self-fulfilling prophecy of a Chinese threat of their own design. There may well be enough threat of Chinese design without the United States exacerbating it.

Hence within the following brief discussion of contemporary issues concerning China an effort must be made to distinguish popular U.S. perceptions, versus Chinese perceptions. Changes within China subsequent to the modernization program have been dramatic in some areas, nonexistent in others, and somewhere in between on the bulk. One thing that can be said with great certainty, however, is that changes have not been fast enough or broad enough for the U.S. press or Congress. Chinese leaders feel that picking friends and basing policy on ideology, as the United States does, is a dangerous mode of operation. China holds that countries ought to seek out areas of commonality and work together on them for the benefit of each, and leave the rest to be dealt with as issues of national sovereignty. The U.S. position is that there are minimum international standards of behavior which must be adhered to in order to be accepted in the Family of Nations, or at least the favored U.S. circle. Each is understandably self-serving. The task for policy analysts, therefore, is to identify paths beneficial to pursue from both perspectives.

Human Rights

The United States has expressed outrage, primarily through politicians running for reelection or Hollywood actors (two groups which have traditionally felt that any press is good press), at Chinese human rights violations in Tibet, in conjunction with the plight of individual dissidents, and most recently concerning persecutions of Christians.[31] Sometimes actions of those in the United States, well intentioned or not, hinder rather than help those in China they seek to assist.[32] The atrocities committed in Tibet, religious intolerance and lack of *habeus corpus* laws akin to those in the United States are certainly abhorrent. That they so ardently grip the attention of influentials and guide the decision making of policy-makers in the United States is, however, surprising to some Chinese.

From the Chinese perspective, the United States in its early years had its share of problems too. Developing countries, as anyone who has ever visited China will attest it is economically, have historically been unable to politically, financially, or economically afford the same standards as developed countries.[33] In the developing days of the United States, indentured servants were commonplace, slaves were used to harvest cotton crops in the South, children were employed in Northern factories, the Indians were run off their land and virtually annihilated, and all in the name of development. But, one could argue, that was a long time ago. Kent State, the Los Angeles riots, and Waco were not, however, and all were human rights violations from somebody's perspective. Further, the Chinese question how Chinese policies are any more abusive than those against the Palestinians or within Saudi Arabia, and what the difference is except that Israel and the Saudis are strategic allies of the United States. American policy is at the least inconsistent, which then makes criticism from the United States difficult for the Chinese to accept or take seriously as coming from "the moral high ground." However confusing though, the fact is that China must contend with what appears to them as prejudicial decision making about human rights issues in their own country and selective procrastination from the United States toward other countries' similar problems.

Chinese citizens understand that they have been recently granted expanded personal freedoms as opposed to *rights.* They also understand that if taken too far, there can still be an ominous knock at the door in the middle of the night. They also see divisions within the dissident groups, rather than seeing "dissidents" as a concordant group all working toward an identical goal. For the most part though, they are so grateful for the significantly higher standard of living that has been achieved that they are not looking to push harder, but want to maintain what they have gained and hope that expansion will inherently occur as a matter of spillover from economic progress. Hence, the average citizen in China is not particularly sympathetic to the plight of an individual dissident and sometimes even sees those individuals as risking the progress that has been made for everyone.

In a 1996 interview, State Councilor and Chairman of the powerful State Science and Technology Commission Song Jian responded to a question about a 1995 petition signed by some leading Chinese scientists and intellectuals asking for increased political freedom. "Of course . . . it is an important principle of the government that China should remain stable. That is a concern of many of our people, including scientists. They do not wish what happened in the former Soviet Union to happen in China. On the other hand, I think there is still room for some improvement."[34] As well as typically circumventing a direct criticism of the government, the response also reflects the genuine concern and priority the average Chinese citizen places on internal stability and the willingness to balance stability and rights. Most American do not understand or appreciate how far the Chinese have come in the past 20 years. After two generations of Maoist society, Chinese today are grateful that they can choose a job, rather than being assigned one for life.

When China defeated U.S. attempts to pass a United Nations resolution condemning its human rights record, which France and Germany refused to cosponsor, Secretary of State Madeleine Albright said that the United States "regretted the defeat."[35] Other American officials insisted that the resolution had been a matter of principle. The Chinese delegate, Wu Jianmin, stated the Chinese position.

> Mr. Wu labeled the resolution "an outrageous distortion of China's reality" and said it reflected Western attempts to "dominate China's fate." He said the Chinese people "had no human rights to speak of" before the Communists came to power in 1949, and he dismissed the resolution criticizing China's rights record and its treatment of Tibet as Western "impudence." He said China had followed its own course for 5,000 years and would continue to do so. Mr. Wu said the West had a powerful interest in preserving China's stability, saying that even if 1 percent of his countrymen—12 million people—were to flee, "the prosperity of East Asia would be destroyed overnight."[36]

China continually stresses evolutionary change to preserve stability, and sovereignty, while the United States continues to ask for changes in China appropriate to the current environment and situation in the United States, but perhaps inappropriate for China.

Meanwhile, it seems unacceptable to Americans to simply stand by and watch abuses occur in China (though it seems to be acceptable elsewhere). It was suggested to me in China that perhaps an alternative to dictating internal Chinese policy would be to assist the Chinese with mechanisms essential to development. For example, teaching nonlethal methods of force in China as the United States does in some Latin American countries, as an alternative to running people down with tanks, might be very useful knowledge toward maintaining political stability, a Chinese priority, without killing citizens and subsequently offending U.S. sensibilities.

Stating that Chinese citizens are not fully supportive of individual or even groups of dissidents is not to say that the Chinese people were not affected by the violence ordered by Deng in response to the student protests at Tiananmen Square in 1989. James Miles points out that, conservatively, 5000 civilians were killed or injured at Tiananmen and if that figure is multiplied by the number of their relatives and close friends, one can grasp the numbers of those intimately connected.[37] The Chinese government is acutely aware of those numbers, hence the need to keep economic reform at a pace fast enough to make acceptance of governmental rules the preferred option to confronting the government about political reform or being held accountable for Tiananmen. The Chinese populace seems to put stock in the premise that with economic reform will eventually come political reform. Significant portions of the population do not believe in communism, but they want economic growth, which is currently being provided by the Communist Party. The Chinese are well aware that economic growth does not occur during times of internal strife, and so seek stability so that growth can continue.

Economics

Deng Xiaoping's economic policies can rightly be credited with saving China from economic collapse. Even with the annual growth rate in China dropping from 10% to 7%, the potential for the Chinese market has the international business community drooling to get in on the ground floor. The Chinese are fervently working to get their house in order, for self-serving reasons, but the house is still a house of cards, precariously balanced and highly vulnerable from inside and out. Clearly, their economic modernization cannot be accomplished without foreign capital, yet the Chinese are not without reason to be skeptical of foreigners. The so-called "hundred years of humiliation" from which they are just emerging was to a large extent brought on by the influx of foreign interests into China in the 19th century. That memory is not gone. Now that China has begun to emerge from its cocoon and catch up to the developed world, it is being done with great deliberation so as not to repeat past mistakes.

With internal political stability essential to the survival of the Communist Party and distributed economic growth the key to stability, full employment of the workforce must be a constant priority goal of the Chinese government. With a population of 1.2 billion this is a daunting, perhaps eventually overwhelming task. Even today, statistics belie how little things have changed in some areas of the country. In 1988, the per capita GNP in China was $320 annually, with 62.5% of the workforce engaged in agriculture and fishing. In 1996, the per capita GNP had jumped to $2,660 annually, with about 60% of the workforce employed in agriculture and fishing.[38] Nearly 300 million people in China still live on less than $1 a day.[39] China is clearly facing an increased gap between its urban population, which is gaining ground economically, and those still living in rural areas, where change has been far slower. This has led to a multitude of social problems already, and the Chinese government is pedaling as fast as it can to keep it from leading to political problems as well.

According to Confucian rules of responsibility, families must take care of their own. There is no obligation on the part of the government to individuals, nor to nonfamily members by individuals. As this is in direct contradiction to communism, the hybrid nature of Chinese communism, especially what is vocally professed versus what is actually practiced, is often confusing from many perspectives. Culturally, the Western propensity to help strangers is viewed as particularly peculiar by the Chinese. Indeed the calamities of strangers, like car accidents, are often viewed as mere distractions from daily toils or a source of amusement in China. It would not occur to anyone to help those in need; "charity" extends only to families, but is mandatory within the family. Hence, a large population has always been a factor driving policy concerns from the family and village level, to the national level.

Traditionally, the ideal in China was to have a large family, particularly to have many sons. Children were a sign of prosperity. Indeed there is a folk tale referencing "100 children" which is often depicted in artwork. The idea is that a family which had 100 children was honorable and strong, and able to take care of its own. Mao also supported the idea of large families to feed the communist mill. The reality, however, is that populations must be fed or they become restless and angry, a fact Chinese leadership is acutely cognizant of today.

Until the Communist Revolution, an estimated 90% or more of the Chinese population was considered peasantry, peasantry which lived in appalling rural conditions. Large families were needed to tend the crops. The negative side of big families though derives from the conditions under which they endured. Human life on Earth at the singular level was not regarded particularly highly. There were simply too many people to worry about individuals, particularly girls, who were considered worth less than boys in the prestige race. The worth of daughters was so low that it was not uncommon for them to go unnamed, and simply referred to by their numerical placement among the other girls. Suffering was hence considered as a expected part of life. Many Westerners are appalled by the Chinese custom of eating cats and dogs, and their seeming cruelty to animals of all types. With human suffering an expected part of their own existence though, the Chinese wonder for the concern outsiders give to mere animals.

Population measures designed to slow the population growth were instituted in 1980, each married couple being limited to only one child, with penalties for violators. In some cases, fees could be paid for permission to have a second child. An illustration of how culture impacts Chinese society came about through an interesting conversation I had with a Chinese teacher I met in Shanghai at one of the parks in the city. As an English teacher and newfound capitalist interested in earning some extra money, she was there offering her services as a guide to English-speaking tourists. As she led me through the park we talked for a long time, especially about children since my son was with me and she had a daughter at home about the same age. I said I had heard that if people paid some sum of money to the government they could have a second child, and asked if that were true. She replied that it might be so, but that the dishonor would be so great that few people even think of it. She said she, for example, would never be able to be employed as a teacher again if she had a second child, the shame would be so great. Admittedly though, policy effectiveness has varied between urban and rural areas.

From the peasant perspective, just as the state was moving away from communal farming and back to family farms, family and hence labor force size were being limited. Additionally at that time, the state also began moving from cradle-to-grave care of the population to family responsibility, with reduced family sizes. The peasant population saw these measures as unbalanced, against them. Subsequently the rural birth rates in the early 1980's actually began to climb.

> The state responded . . . with harsher measures to enforce the norm of the one-child family, including confiscation of contracted lands and forced sterilizations and abortions (sometimes as late as the third trimester) in cases where social persuasion and political threats failed. Peasants, in turn, responded with various stratagems to circumvent official birth-control regulations, sometimes, in extremity, resorting to female infanticide and the abandonment of female infants. The tragic irony of this grim battle is that both the Communist state and the peasantry acted in accordance with the dictates of "economic rationality."[40]

Clearly, this is an issue where individual rights versus collective interests comes glaring into play.

The Chinese see the population policy as a draconian edict in relation to their Confucian heritage. But it is one that the current urban generation seems mostly willing to accept as the sacrifice they have to make for China. The still majority peasant population, however, have far less motivation (inherent, rather than forced) to accommodate government policy. In 1985 the policy was relaxed somewhat and has come to unofficially mean that rural families can be allowed two children. In 1997 that was extended, selectively, to some urban areas as well.[41] The other side of the coin though is that the goal of population stabilization in China by 2000 will not likely be achieved.

There has always been a cultural sense of elitism among Chinese city dwellers toward the rural peasant population. Now with an estimated 90–130 million Chinese as a "floating" population, those who have left the countryside for a better life in the city but with nowhere to go and no official papers, there is a marked increase in crime. Approximately 60% of those arrested in property-related crime are first time offenders and 70% of the crime is committed by those considered "dislocated." Parts of south Beijing are considered unsafe, even for local Chinese, because of the high number of "floaters" there. Consequently, what was once urban elitism toward the rural population is now turning toward animosity.

Returning to the goal of stability, a goal shared by the government and increasingly prosperous members of the society, there is intense governmental awareness that too many people cannot be left behind economically without eventual negative impact.[42] Revolution in China has always come from rural discontent. Chinese efforts to link their country through cellular phones are more than noble. Chinese leaders are mindful of the 1949 revolution being led by farmers tired of being led by what they perceived as corrupt city dwellers disconnected from them and their needs. Although 26% of the Chinese population lives in urban areas, telephone density overall in China was 3.76% in

1995, with only 0.49% in the rural areas.[43] The Chinese government needs strong contacts to their interior provinces established as soon as possible.

The precarious position of the central government is exacerbated because of the power of the regional and local leaders in all areas, including taxation. Structurally, one of the more difficult imperial legacies has been the practice of having taxes collected by county magistrates, with no national tax collection agency. This system, not surprisingly, then and today has meant that funds are siphoned off by various percentages all the way through the chain from the village to the national level, with varying levels of corresponding loyalties between those doing the collecting and those doing the distribution. From the local perspective, they say that substantial local initiative is required to deal with the diverse issues faced within the regions. The fact remains, however, that the national government has an unpredictable and unreliable tax base upon which to plan.

Another aspect of economic challenge for the Chinese government is privatizing state-owned enterprises (SOE). Moving to capitalism from an iron rice bowl society involves stages. The first step is setting up the means by which the population can gradually be released from the welfare existence which characterizes an economic system where the government is the sole employer. Although only SOEs are allowed to hold stakes in defense and some high-technology industries, the rest of the former SOEs are being turned loose at what some observers characterize as amazing speed. It must also be understood that in China the government is increasingly becoming the largest entrepreneur of all. Government funding to previously supported activities from factories to schools to the army is shrinking, with simultaneous encouragement to *xiahai*, or "down to the sea," meaning to go into business for themselves.[44] For example, the Ministry of Public Security (internal secret police) owns luxury hotels and the State Security Ministry (foreign and counter intelligence) operates an import-export company. The PLA is perhaps learning capitalism fastest of all. It owns interests in businesses including luxury hotels, converted weapons factories now producing bicycles, television factories and, not surprising, arms exports. It is estimated that the PLA is involved in over 20,000 entrepreneurial ventures.[45] Not only does the PLA support itself through domestic cash-generating enterprises, foreign ventures are involved as well. A former vice president at Chase Manhattan bank, Roger Robinson, has uncovered more than $6 billion in Chinese-issued bonds being offered in the United States, bonds with ties to the PLA and the Chinese military-industrial complex.[46]

Another aspect of the gradual breaking of the iron rice bowl is that in this interim stage many Chinese prospering from the economic reform retain many of the benefits a state-run society offered. While perhaps earning increased wages or generating their own profit, housing, medical, and education costs are still paid for or heavily supplemented by the government. This increases the amount of disposable income available to this group by a significant amount. They are characterized as akin to teens with an allowance who can still count on mom and dad to provide the basics. Eventually, however, the birds will get kicked from the nest, and the Chinese government needs to be sure that they are prepared.

Foreign and Defense Policy

In terms of actual military capabilities, analysts disagree about China. Some feel that the increase in the PLA budget, the acquisition of the Su-27 fighter aircraft from Russia, and the expanding of the Chinese navy are all evidence that China is seeking to establish hegemony over the region. Others maintain that China's military modernization program is limited in scope and insufficient for sustained, extended power projection operations. Still others argue that regional apprehension over Chinese military modernization is the result of traditional anxiety over the ambiguity of Beijing's intentions. Finally, another group says that alarm being voiced by other countries, and not necessarily just in the immediate region, is actually part of their own efforts to justify and increase military spending levels.[47]

Modernization of the Chinese military clearly has not been their number one priority. The U.S. rationale for this low prioritization has been that the U.S. presence in the region acted as a stabilizing effect, on the Japanese and Koreans in particular, allowing China to forego military modernization until the economic outlook improves further. Although the Chinese had to officially reject the notion of desiring U.S. troops in North Asia, unofficially and privately there was some recognition of the benefits to the Chinese. Recently, however, the Chinese seem to be changing their view of that premise. Indeed with increasingly regularity Chinese claims are made that the need for military modernization is intensifying because of U.S. interference in the Taiwan issue.

The view from Beijing is that the Cold War is over and it is time to redefine relationships. The Chinese feel, however, that the United States is still defining relationships on Cold War terms. Clearly, the Chinese want and need stability in North Asia. Increasingly, however, they do not see a U.S. presence as favorable toward achieving that goal. The United States, per their view, has gone out of its way to inhibit Chinese recovery from a self-perceived century of shame.

"Face" is important in the Chinese culture, as it is in every Asian culture. Symbolism is equal to or more important than substance. In that regard, the Chinese are seeking international recognition of having shaken off a bad couple of centuries. They bid for and badly wanted to host the 2000 Olympics in Beijing as a coming out party to which all were to be invited. The Chinese fault the Americans for having lost the competition to Sydney, Australia, by the United States lending its support to the latter.

The Chinese are now seeking admission to the WTO, only to see their membership being blocked by the United States. China cannot afford to join unless it is allowed membership as a developing country. The United States feels China ought to be admitted as a developed country. Entering the WTO is a clear goal of the Chinese, but one "that doesn't justify wholesale job losses or weakening its industrial base,"[48] which the Chinese see entering under U.S. terms as having the potential of doing.

The two issues on which the United States and China clash most, however, are Taiwan and Hong Kong. Just as one man's terrorist is another man's freedom fighter, to China, Taiwan is a renegade province, to the United States, it is a fledgling democracy. Just as the United States knows that China does not currently have the military capability to invade Taiwan, Taiwan also knows that China could lower the economic hammer any day it chose, and Taiwan is a island captivated by commerce as much or more as it is democratic ideals.[49] So when China scheduled its missile tests over the Formosa Straits in March 1996 just prior to the Taiwanese elections, the immediate military intent (nothing) was clear. It was a symbolic act on the part of the mainland to remind Taiwan not to stray too far from home. Likely it was as much for internal posturing for prominence and funding by PLA as for a signal to Taiwan. If the United States wanted to object prior, the options were several, including sending the naval carriers to the region as it did eventually. To wait until the exercise commenced, however, surprised the Chinese and caused them to lose face on what they viewed as an internal matter. According to the Chinese, the United States sees Taiwan as an unsinkable aircraft carrier for its exclusive use. Also, that the issue gives U.S. Congressional hawks a good excuse for their pleas for more funding for ballistic missile money (Thaad), Navy theater wide (upper tier) defense, and theater missile defense (TMD) is not lost on the Chinese.

Hong Kong is another issue on which the United States and Chinese have widely different perspectives. Although recognizing that the United States and other countries have substantial business interests in Hong Kong, the Chinese are adamant that Hong Kong is part of China and hence the United States has no business telling the Chinese how it ought to be run. The analogy given is the following: imagine that the British had kept Manhattan Island after the Revolutionary War, to be returned in 1997. What business would it be of the French, Germans, Japanese, or anyone else how the United States intended to then govern what was acknowledgedly part of the United States? Further, by so openly and negatively speculating about difficulties anticipated when Hong Kong reverted, China says that the United States did its best to deliberately poison the public opinion well with preconceived notions based on the old Cold War ideologies again. They pointed out that even the traditionally dissident Hong Kong press toned down its pre-reversion rhetoric in support of "balance" and making the transition smoother,[50] but that the United States was unwilling to do the same.

The Chinese were anxious to welcome Hong Kong back to its embrace for a number of reasons, many pragmatic as well as nationalistic. Hong Kong might well serve as the money machine to push the rest of China further down the very long road of development it faces. Hence the importance of Hong Kong to the mainland gives Hong Kong itself a certain amount of leverage, as does the international attention it will receive. Taiwan, for example, will be watching closely to see how Hong Kong is integrated, and this could well influence its future decision making.

Clearly, the integration will not be without "challenges." Hong Kong has a viable legal system based on common law. It is especially mature in areas of international commercial law around which Hong Kong has long built its livelihood. This as opposed to the virtual legal void in commercial law found throughout the rest of China. Further, Hong Kong has approximately 600 laws concerned with governance, and China let it be known well before 1 July 1997 that it wanted to change approximately 25 of them. Of those 25, some were archaic, some dealt with elections, and some with civil liberties. Clearly it is the latter two categories which have been problematic. In April 1997 the Chinese announced plans to restrict public protest and rights of association in the name of social stability. As stability is Beijing's number one priority, the measures were not surprising, but still met with harsh rebukes within Hong Kong.[51] The government and populace of Hong Kong wanted Hong Kong to remain virtually unchanged after the 1 July reversion. They knew, however, that was unrealistic and began to accommodate and compromise accordingly.

Hong Kong is also something of a test bed for the rest of China concerning how far a legal system can actually inhibit the growing influence of corruption there. The turmoil which hit the stock markets in October 1997, starting in Hong Kong and reverberating throughout Southeast Asia and beyond, may signal cracks in the "one country, two systems" approach which Beijing promised would prevail in Hong Kong after the takeover. The bubble economy which had developed in Hong Kong was sure to burst, as bubbles always do, but those negatively impacted won't quickly forget the financial sting. The reasons behind the turmoil involve both the greed of investors racing to invest in "Red Chip" stocks, including space-related ones like China Telecom which had risen in recent months and showed the potential for even more lucrative increases, and the willingness of the Chinese to provide opportunities for investment, even if only in "shell" companies.[52] The role of the Princelings is still unknown, or at least unmeasured. Whether the impact of the Princelings/Cadre Kids can be limited through the courts will be closely watched by everyone, especially those within China weary of their parasitic manner. The first question on that issue, however, is whether anyone will have the courage to challenge the Princelings in the legal system. If they are challenged and the courts are ineffectual, that might be the shot heard round the world to the business community.

The Chinese space program will be subject to the same influences which have been described in this chapter as influencing and guiding other aspects of contemporary Chinese politics. Space is not an end unto itself in China, but rather a means to a greater "end." Although that is true in other countries as well, the degree to which it is the case in China is exacerbated. Therefore, it is critical to understand what the desired macro "end" is, where it derives from, competing and complimentary secondary goals, and the parameters within which the Chinese (as opposed to Americans or others) are willing to operate or use the available "means" to achieve their goals.

Conclusions

In his 1969 book *To Change China,* Jonathan Spence analyzed efforts made between 1620–1960 by Western advisors to convert China to Western ways by proving the superiority of Western science. With the successful testing of the Chinese H-bomb in June 1967, China-the-pupil seemed to have mastered the lessons of the teachers. But, Spence concludes:

> ... China had not been converted, in any of the ways that generations of Western advisers had hoped. Her land was not dotted with the steeples of Christian churches, no Chinese senators rose in marble halls to extoll the virtues of the "democratic way of life," on the benches of her schools no eager students were noting the gems of Western humanism or applauding the current practice of Soviet communism. For China there were no values implicit in these images, only the bitter flavor of exploitation, deceit, and betrayal; or, as they phrased it, of imperialism, bourgeois individualism, and revisionism. To most Westerners, China's exterior was again as forbidding as it had been in the late Ming dynasty, and her internal workings as mysterious.[53]

Clearly, those assuming that analyzing or interacting with China is the same as dealing with any other country are deluding themselves. Yet the tendency has been to act as though that were true. A first step then in appropriately and adequately dealing with China is to at least attempt to know the Chinese better. If lessons can be learned from the past, then Chinese history is replete with insight to offer Western analysts.

Some analytical and interactive lessons seem clearer than others. Whatever the issue, analysts ranging from the congenital conservatives to the State Department officials considered to have "gone native" after being in a country too long, agree that the United States needs to send clearer signals to the Chinese about U.S. interests and intentions. China is a rising power. As such it is both insecure and overly confident. Rising powers have been traditionally difficult to deal with, Germany pre-World War I being an example. China is at a point where it wants to reclaim what it considers to be legitimately its own, Hong Kong and Taiwan being cases in point, though questions to its claim may be seen as legitimate outside China. There are others areas which could erupt into hot spots in the future, the Spratly Islands being one such example.

The Chinese government has a full set of issues to deal with in the near future. With all the balls that it must effectively juggle in the air, however, there seems to be clear recognition that there can only be one number one priority. There are a multitude of "important" issues, but the number one priority is internal stability, which can be bought from the populace with economic development. This buys the leadership time to transition in ways which still have yet to be fully decided, so overwhelming is the stability goal. The lessons of history generally tell us that economic development is most often purchased at a price to civil liberties; Chinese history tells us that that price is not considered exorbitant.

Within that context, the Chinese will likely posture to give themselves maximum flexibility in the future, when the internal dynamics of China actually allow choices to be made based on other than the currently overriding issue of steadiness. That inherently requires that the status quo will be supported whenever and wherever possible in terms of decision-making processes. That also means that the Chinese will push where they feel they will encounter little resistance, and avoid areas of confrontation, at home and abroad. Therefore, the United States and other countries need to carefully consider their own position regarding China, taking opportunities to provide prescient tutorials where asked or allowed, provide clear guidance about U.S. policy and interests, and when tempted to preach from a glass house, resist.

Endnotes

1. Alain Peyrefitte, *The Chinese, Portrait of a People,* translated from the French by Graham Webb, (New York: Bobbs-Merrill Company, Inc., 1973) 244.
2. Those used as references for this work include: Mark Elvin, *The Pattern of the China Past,* (Stanford, CA: Stanford University Press, 1973); Kenneth Lieberthal, *Governing China,* (New York: W. W. Norton & Company, 1995); L. Carrington Goodrich, *A Short History of the Chinese People,* (New York: Harper & Row, 1943); Stephen G. Haw, *A Traveler's History of China,* (New York: Interlink Books, 1995).
3. Peyrefitte, 244.
4. *U.S. and Asia Statistical Handbook 1996,* Compiled by John T. Dori and Richard D. Fisher, Jr., (The Heritage Foundation, Asian Studies Center) 38.
5. L. Carrington Goodrich, *A Short History of the Chinese People,* (New York: Harper & Brothers, 1943) 192–195.
6. Stephen G. Haw, 147.
7. Former Premier Zhou Enlai was another exception. He studied in both Japan and

Europe. Later, it was Zhou who was the architect of the new relationship between China and the United States in the early 1970's.

8. Yan Jiaqi and Gao Gao, *Turbulent Decade: A History of the Cultural Revolution,* (Honolulu: University of Hawai'i Press, 1996) 1.
9. Roger Garside, *Coming Alive: China After Mao,* (New York: McGraw-Hill Book Company, 1981) 54.
10. Maurice Meisner, *The Deng Xiaoping Era,* (New York: Hill and Wang, 1996) 67.
11. Joan Johnson-Freese and Roger Handberg, *Space, the Dormant Frontier: Changing the Space Paradigm for the 21st Century,* (Westport, CT: Praeger Books, 1997).
12. Fred W. Riggs, "The Theory of Political Development," *Contemporary Political Analysis,* ed. James C. Charlesworth (New York: The Free Press, 1967) 317–349.
13. Fred Riggs, 341.
14. Graham T. Allison, *The American Political Science Review,* LXIII, 3, September 1969, 689–718.
15. *A National Security Strategy for a New Century,* The White House, May 1997, 19.
16. (New York: W. W. Norton & Company, 1995).
17. Lieberthal, xv.
18. Morton A. Kaplan, "Systems Theory," in *Contemporary Political Analysis*, James C. Charlesworth, ed., (New York: Free Press, 1967) 150–163.
19. Lieberthal, xvi.
20. "China's 'Last Emperor,' " *The Washington Post,* 20 February 1997, A26.
21. As evidenced by the relatively quick ousting of two of Jiang's leading critics. Steven Mufson, "Chinese Shake-Up Leadership, *The Washington Post,* 19 September 1997.
22. Kathy Chen, "After Deng: A Look at the Men Who Will Run China," *The Wall Street Journal,* 20 February 1997, A17.
23. Mark L. Clifford, et al., "Can China Reform Its Economy?" *Business Week*, 29 September 1997, 119.
24. *A National Security Strategy of Engagement and Enlargement,* The White House, February 1996.
25. Joseph S. Nye, Jr., "An Engaging China Policy," *The Wall Street Journal,* 13 May 1997,
26. Robert S. Greenberger, "Favored-Nation Status for China Loses Its Certainty," *The Wall Street Journal,* 15 April 1997, A20; Jeffrey Taylor, "China Puts Heat on U.S. Firms to Lobby for 'Most-Favored-Nation' Trade Status," *The Wall Street Journal,* 24 June 1997, A24;
27. Robert S. Leiken, "Controlling the Global Corruption Epidemic," *Foreign Policy,* Winter 1996–96, 55–76.
28. Robert Greenberger and David Rogers, "China Vote Divides GOP's Christian and Business Wings," *The Wall Street Journal,* 24 June 1997, A24; Paul Blustein, "The China Syndrome: Despite the criticism, engaging Beijing may be the United States' only option," *Washington Post National Weekly Edition,* 21 April 1997, 23.
29. *The APT Yearbook,* 1997, A. Narayan, ed., (Surrey, U.K.: ICOM Publications Ltd., 1997) 307.
30. *The APT Yearbook,* 1997, A. Narayan, ed., (Surrey, U.K.: ICOM Publications Ltd., 1997) 307.
31. C. Bogert, "Pray for China," *Newsweek,* 9 June 1997, 44–45.
32. Stan Crock, "The Wrong Way To Strike At Religious Persecution," *Business Week*, 29 September 1997, 42.
33. One of the aspects of China which most surprised me, and many other Americans I

have spoken with, is the unbelievable pollution. The Sun never broke through some days in Beijing, Shanghai, and Chengdu, creating an almost claustrophobic feeling. The reason is obvious though. Trucks filled with dirty soft coal line the streets, the coal sold for heating and power generation. Developing countries use what they can get and what they can afford, not what is environmentally friendly. Environmental laws were a luxury that could not be afforded in the United States until the 1960's and 1970's, so it is no wonder that they are missing in China today. Even using coal, brownouts are not uncommon in many Chinese cities, with power generation unable to keep up with power needs.

34. Fisher, 42.
35. "China Defeats a U.N. Resolution Criticizing Its Human Rights Record," *The New York Times,* 16 April 1997, A11.
36. Ibid.
37. Miles, 16. Over 55,000 mourners turned out in Hong Kong in what they considered perhaps the last Tiananmen memorial vigil pre-handover. Chris Yeung, et al., "Mourners Struggle to the End," *South China Morning Post,* 5 June 1997, 1.
38. See: *U.S. and Asia Statistical Handbook 1996,* Compiled and edited by John T. Dori and Richard D. Fisher, Jr., (The Heritage Foundation, Asian Studies Center) 38,39; *U.S. and Asia Statistical Handbook 1989,* Compiled and edited by Thomas J. Timmons (The Heritage Foundation, Asian Studies Center) 26,27.
39. Clifford, 119.
40. Meisner, 244.
41. Joseph Kahn, "China Eases Up on Its One-Child Policy," *The Wall Street Journal,* 20 October 1997, A20.
42. "China's Growing Inequality: As Economy Takes Off, Millions Are Left Behind," *The Washington Post,* 1 January 1997, A1.
43. *The APT Yearbook,* 1997, 307.
44. See, for example, Godfrey Kwok-yung Yeung, "The People's Liberation Army and the Market Economy," *China in the 1990's,* ed. Robert Benewick and Paul Wingrove, (Vancouver: UBC Press, 1995) 158-168.
45. Meisner, 334.
46. Peter Schweitzer, "You, Too, May Be Funding China's Army," *USA Today,* 14 May 1997, A13; Solomon M. Karmel, "The Chinese Military's Hunt for Profits," *Foreign Policy,* Summer 1997, 102-112.
47. John Caldwell, Col. (USMC), *China's Conventional Military Capabilities, 1994-2004,* Center for Strategic and International Studies, 1.
48. Eduardo Lachica, "Chinese Official Says WTO Pace May Take Time," *The Wall Street Journal,* 18 April 1997.
49. Ian Buruma, "Taiwan's New Nationalists," *Foreign Affairs,* July/August 1996,
50. Joseph Kahn, "China Has No Need to Suppress the Press in Hong Kong Now," *The Wall Street Journal,* 21 April 1997, A1.
51. Edward A. Gargan, "Hong Kong Erupts at Plan to Curtain Rights," *The New York Times,* 11 April 1997, A8.
52. Marcus W. Brauchli, Craig S. Smith, and Joseph Kahn, "Asian Market Turmoil Adds 'Safe' Hong Kong to Its List of Victims," *The Wall Street Journal,* 24 October 1997, A1, A19.
53. Jonathan Spence, *To Change China,* (Boston: Little, Brown, 1969) 289.

Chapter 3

China's History in Space

Introduction

China's extensive history is replete with examples of its technological innovations. Gunpowder, printing, harnesses, and stirrups are all attributed to Chinese ingenuity. Indeed China has been working on rocket technology for some 800 years. But China's technological history is one characterized by periods of great advancement, followed by not just stagnation, but often destruction. China failed to undergo a Galilean revolution such as catapulted Europe forward into the age of modern science in the 17th century, and subsequently failed to transition from an agrarian to an industrial society. Much of the reason for that failure is attributed to the absence of a merchant class and the presence of a suffocating bureaucracy in its place. The Cultural Revolution in the 1960's obliterated the advancements that had been made in China after its self-imposed isolation. Space-related fields did not escape the patterns of the distant past nor the wrath of the 1960's. The history of China's space program is somewhat exegetic.

Cold War Roots

As in the United States and the Soviet Union, the Chinese space program was founded not in the hopes of exploring the Heavens or even for the more mundane and pragmatic goal of economic profit, but rather as part of their Cold War defense policy. Indeed, the Chinese themselves say "Especially the development of the ballistic surface to surface missiles laid a foundation for the development of space launch vehicles."[1] The technological aspects of defense policy in the post-World War II years focused on parallel emphases on nuclear weapons development and delivery systems for both the Soviet Union and China. In the United States, with those capabilities available via the wartime Manhattan Project and long-range bombers, the emphasis was originally on surveillance into the closed communist societies via satellites and missiles to launch them, and later missiles to act as a delivery systems as part of the strategic triad.[2] In China, the weapons research and missile research was in many ways juxtaposed and singular, a strategic program. The priority of the overall strategic program, during some particularly difficult years in China, was such that it was "a magnitude equal to America's Manhattan Project and postwar missile program combined."[3]

In 1956, Soviet advisors in Beijing strongly recommended the inclusion of missile technology in the 12-year plan for scientific and technical development then under development to cover 1956–1967.[4] About the same time, Qian Xuesen, a Chinese rocket specialist with a doctorate from the California Institute of Technology who had participated in an American military survey of the German missile industry after World War II and indeed worked on the early research and development (R&D) of U.S. missiles, was returned to China after receiving considerable notoriety. Qian, who had once sought U.S. citizenship, was a victim of the McCarthy era, accused of being a Communist despite lack of evidence against him. Held under virtual house arrest for 5 years and then deported, he quickly became one of the most powerful scientists in China and retained, not surprisingly, considerable mistrust toward the United States as the country which had turned against him.[5] He wrote a proactive article on missile development which became a proposal to the Chinese leadership.

From the leadership perspective, following through on the cognate recommendations of both the Soviets and Qian and his associates would fill a defense void having to do with the Chinese world view of America as the enemy, and an enemy which had repeatedly threatened China with nuclear attacks. China understood that only long-range ballistic missiles could reach the homeland of the United States, and set off on a path toward acquiring those missiles. From that point until the early 1980's China was reacting to perceived U.S. threats from a technological capability perspective, and to Soviet threats as well, rather than through any concerted strategic planning or even tactical consideration on its part.

That it was the Communist Party which organizationally responded to the quest for missiles indicates that these were decisions made at the highest levels of government. The party's Central Military Commission created what was designated as the Fifth Academy as an R&D organization within the Defense Ministry. In 1956, they were assigned overall responsibility for building the missiles to respond to the U.S. threat to China.

In terms of the big picture, organizationally from 1956 to 1958 Marshal Nie Rongzhen coordinated China's strategic activity, under the presumed direction of the Central Military Commission and the State Council. The chaos which ensued in the fall of 1957 following the Great Leap Forward policy initiation, however, caused considerable difficulty in program implementation.

Subsequently in 1958, Nie convinced the party leadership to consolidate coordination of the strategic program through the newly formed National Defense Science, Technology and Industry Commission (NDSTIC),[6] which he was to head.

Initially, the Fifth Academy (which became part of NDSTIC in 1959) first began its work with assistance from the Soviet Union. Two R-1 missiles and their documentation were sent to Beijing in September 1956. These were basically copies of V-2s and of limited utility. It was not until Khruschev needed the support of Beijing against political opponents in Eastern Europe and indeed Moscow, that he sent two more sophisticated R-2 missiles in 1957. The Chinese called this the 1059 program, though it was dubbed the SS-2 "Sibling" missile in the West.[7] The missiles arrived at the Fifth Academy by train in the middle of the night. These were followed by the delivery of over 10,000 volumes of blueprints and technical documents for manufacturing, testing, and launching the R-2s in the second half of 1958. This gave the Chinese their first real opportunity to work with an operational missile system.[8]

After the launch of Sputnik in 1957, at the Second Session of the Eighth National Congress of the Communist Party of China in May 1958, Mao Zedong stated "We want to make artificial satellites."[9] Nie Rongzhen immediately added satellites to the strategic program. The Chinese Academy of Sciences and the Fifth Academy began working on a satellite program, with Qian as the group leader. Hence the third aspect of Chinese space activity was also underway.

The years between 1960–1962 are known in China as the "three hard years." The policies of the Great Leap Forward resulted in severe famine and hardship for the Chinese people, people for whom hunger and suffering were the norm. The military and the defense sector did not go unaffected. That the strategic programs were all military or quasi-military programs is clear. Even the industrial sector involved was under strict military control; those working on the missile programs even wore PLA uniforms until 1965. Not only was malnutrition rampant among those working on the strategic program, but the political competition for funds between the strategic programs and those seeking to fortify the conventional forces was heated. Even those programs of the highest priority were impacted, and those of even slightly less priority were seriously inhibited. Indeed in January 1959, General Secretary Deng Xiaoping stated that the national capabilities did not immediately allow the launching of an artificial satellite.[10] Work was slowed on the program, although a sounding rocket was launched in 1960.

Again, between 1958 and 1964 the Chinese efforts were organized toward not just obtaining the weapons capability but obtaining an appropriate delivery system for the weapon built as well. The feeling in China was that "we knew the bomb would work; it was the missile program that was the unknown for us."[11] It quickly became clear that even the R-2 could not reach American bases in Japan with its 590 km range, and that the atomic warhead under development would exceed the payload capacity of the vehicle also. Therefore, efforts began on another program as well.

In 1958, along with the 1059 program (by then including limited access to the larger R-5/SS-3 "Shyster" missile), the Chinese instituted what they referred to as the Dong Feng (DF, or East wind) program of land-based missile development. The name references a speech made by Mao in 1957 in which he said "the East wind prevails over the West wind" which became popular and immured in all aspects of everyday life.

Professional and private life for those working on the Chinese space program at this time was spartan at best, more often crude and even dangerous. Engineers lived and ate in communal facilities, doing their own laundry and hanging it to dry in their rooms. The "command post" at a makeshift launch site near Shanghai used for testing consisted of a heap of sandbags piled to protect observers during a launch, with orders shouted between the launch commander and those unlucky workers who acted as troubleshooters and hence had a proximity to the rocket at launch that was often perilous. A bicycle pump was used to load the rocket propellant.[12]

Then in 1960, Chinese engineers basically realized they were on their own with the break with the Soviet Union. Mao had shocked Khruschev in 1957 with his welcoming comment about the prospect of nuclear war: "We may lose more than three hundred million people. So what? War is war. The years will pass and we'll get to work producing more babies."[13] Khruschev quickly decided that Mao was a madman of the Stalin variety, and that he was not going to help him get the means with which to blow up the world.[14] Beginning to pull back on the Soviet promise to help the Chinese build nuclear weapons almost as soon as it was made in 1957, the final Sino-Soviet split was then just a matter of time.

The Chinese then in 1960 took the 1059 program and began working on an evolved DF-2 vehicle from it, a vehicle with a range capable of reaching Japan. The real target, however, was the United States, and so in 1961 the DF-3 (first called the DF-1) Inter-Continental Ballistic Missile (ICBM) program was initiated. Qian Xuesen himself was to be the chief designer. Domestic Chinese politics reared it head at this point, however, suffering its technology programs a severe blow.

> This projected ICBM was called DF-3. Soon thereafter, technical setbacks in the missile program and the national economic crisis produced by the politically-motivated Great Leap Forward introduced the cooling winds of reality. Developing a missile capable of reaching North America would have to be accomplished step by step, and in 1963 the Fifth Academy canceled the ICBM version of the DF-3.[15]

Thereafter, the DF-3 became an Intermediate Range Ballistic Missile (IRBM) program, with range and payload capacity capable of targeting U.S. bases in the Philippines.

The Chinese at this time adapted a more deliberate incremental approach to reaching, literally, its strategic targets. In 1964 a series of meetings were held in which a logical sequence was developed for the DF missiles: the DF-2 would be capable of hitting Japan, the DF-3 the Philippines, the DF-4 Guam, and the DF-5 the continental United States. At the same time a philosophical debate arose concerning the concept plans for missile guidance and control. Qian, from the American school, favored an approach which placed heavy emphasis and reliance on individual, sophisticated pieces of equipment. Other engineers more influenced by the Soviets, however, favored being less concerned about the quality of individual components than with how the rocket worked as a system. A compromise was reached which said that the shorter-range missiles, the DF-2, DF-3, and DF-4, would use primitive strap-on accelerometers, but that the DF-5 would use advanced gyroscopes with stabilized platforms and gimbals installed inside the rockets. Qian was said to have insisted on the latter, not wanting to "be satisfied with a primitive intercontinental ballistic missile."[16]

It was not until after the break between Beijing and Moscow that the Chinese converted its existing production base to produce its first indigenous missile, the CSS-1 (Chinese surface-to-surface missile number 1). Although the CSS-1 served well as a transition from Soviet to indigenous designs, the CSS-1 was of limited value as a weapon. Believed to use liquid oxygen, it could not be stored fully fueled, or hence launched within a short period of time. As short notice launch is a requirement of any operational missile system, the Chinese were forced to move forward to the development of the CSS-2 in the early 1960's. This single-stage missile utilized storable propellant, and was capable of carrying a 2-ton nuclear warhead 2800 kilometers. The CSS-2 became common stock in the Chinese arsenal of intermediate range ballistic missiles. The range of the CSS-2 limited its targets, especially those in the interior of the Soviet Union. That limitation led to the parallel development of the CSS-3, with a similar payload capacity to the CSS-2, but a increased range of about 6500 kilometers. It is from the CSS-3 that the Chinese civil applications launcher series, the Long March rocket,[17] derives.

In November 1965, the Chinese launched the DF-2A, an improved version of the (to that date failed) DF-2 and the country's first inertially guided missile. With it the Chinese also moved from utilizing radar control of the rocket (whereby an enemy could change the direction of the rocket by interfering with its radio signals) to use of a primitive computer placed inside the body of the rocket. In one of the most dangerous experiments in history, in October 1966 a DF-2A rocket was tested, carrying an atomic bomb also being tested![18]

The Cultural Revolution Years

The Cultural Revolution had a major destructive impact on technological advancement in China, with space being hard hit. Examples abound, including "... R&D on the DF-5 proceeded in parallel with that of the DF-4, and both programs were heavily influenced by the actions of radical elements during the Cultural Revolution."[19] It is actually surprising that much was accomplished at all with the persecution and out-and-out fighting that took place. Other examples point out the wide variety of destruction that reigned, from a personal to an institutional level.

Xie Xide is a revered member of China's elite space elders. Educated in the United States and an avid Red Sox fan since her days at MIT, she was president of Shanghai's prestigious Fudan University from 1983 to 1988 and is still a professor of physics there. She returned to China in 1952 to assist her fellow countrymen in establishing their laboratories.

> ... we were making real progress on the eve of the Cultural Revolution. Then, beginning in 1966, everything stopped. Like all other intellectuals, I had to be "criticized," and I lost the 10 best years of my career. The universities began to accept students again in 1971, but not according to their merits, because there was no examination. They were so-called worker-peasant-soldier students, and most were hopelessly unqualified. Besides, much basic research was deemed to be useless and was suspended. So in 1976 we had to start all over again.[20]

Her story was not unique, and is told in many variations.

Prior to the Cultural Revolution, the Chinese were quite advanced in payload work. But, to give another example of the negative impact of the Cultural Revolution on the space program, during the Cultural Revolution the Red Guard particularly targeted the engineers at the 504 Institute at Xi'an, the institute responsible for payloads. The Guard used dynamite against the Institute, the engineers fought back with more advanced plastic explosives. During the assembly of one such device an accident occurred and wiped out a primary group of payload specialists.

The entire Seventh Ministry[21] initially factionalized during the Cultural Revolution, culminating with one of the young engineers (the son of the famous Chinese general Ye Zhengguang) successfully engineering a coup to overthrow Qian and others from power. Qian's status became that of an in-house exile for many months. Violence and distrust permeated the Ministry, which was in virtual disarray. In June 1968 a distinguished metallurgist and head of the 703rd Institute (Beijing Research Institute of Materials Technology), Yau Tongbin, was beaten to death by Red Guards. This incident prompted Zhou Enlai to issue orders protecting scientists and engineers in space and missile research institutes as the programs they supported were high on the list of governmental priorities but were becoming seriously jeopardized by the violence. Qian became one of less than a hundred scientists who received personal protection from the state in Beijing.[22] He thereafter was not only physically protected but became politically powerful, though not immune to the indignities of the Cultural Revolution.

Amazingly, the launch of the first Chinese satellite was still accomplished during this time of incredible personal and institutional struggle. The satellite, the Dong Fang Hong-1,[23] was launched on a Long March 1 (LM-1) vehicle in 1970. Subsequently in 1971, China successfully launched a scientific satellite, called Shi Jian-1 (SJ-1), also on an LM-1. This was China's first experiment in the technology of a long-life satellite. Solar cells were utilized for power on the satellite, as well as advanced active thermal control technology. The various instruments on board operated for 8 years in orbit.

In the late 1960's, China had also initiated research and development program for an experimental recoverable satellite. A compatible launch vehicle was also needed, prompting the LM-2 program. The symbiotic nature of the technology and the intertwined nature of the Chinese military and civil space program allowed parallel development of a missile/launcher for dual application purposes. The project reached fruition with the first Chinese launch of a recoverable satellite on 26 November 1975.

Recovery: 1976–1986

With China's vast territory and a topography which is 80% mountainous land or desert which make land-based long-distance communication difficult, the potential utility of satellite applications quickly became apparent. In the early 1970's, a communication satellite project was initiated. In parallel, efforts began on terrestrial station development and satellite communication technique research.[24] In 1978, the Franco-German Symphonie satellite was used by China to conduct transmission tests for telephone, television, facsimile, time-synchronization, and teleconferencing. These transmission tests were expanded in 1982, using an Intelsat transponder. These programs acted as precursors to a high priority effort supported by the Chinese government which commenced in 1985.

Continuing to pursue its defense objectives, the next step in Chinese missile development was a full-size ICBM, using storable propellants, with a range in excess of 11,000 kilometers. "The fact that this missile [referred to as both the CSS-4 and DF-5, depending on the source] did not have its initial launch until 1979 and was in fact preceded by derivatives used as satellite launchers before fulfilling its original mission as an ICBM has been the cause of considerable speculation."[25] Two separate design teams were established "to develop the CSS-4-based satellite launch vehicle to a common specification."[26] The China Academy of Launch Vehicle Technology, aka First Academy in

Beijing, worked on the one version (nonmilitary), while the Shanghai Bureau of Astronautics, aka Eighth Academy, worked on the other (military). Apparently, Jiang Qing (Madame Mao) insisted that her hometown, Shanghai, be established as a space research and development center as well as Beijing. This fit well with Mao's strategic requirements of protecting key projects by having multiple centers of industry.

Military application had first priority, with the vehicle developed called the Feng Bao ("Storm," FB-1). It was subsequently used several times throughout the 1970's, but suffered from a poor reliability record of about four out of seven successful launches. Possible failures include the first two, in 1973 and 1974.

The nonmilitary version, the LM-2, had far greater success. The LM-2 carries a payload of 2.4 tons. Although the first launch (LM-2A) in 1974 was a failure, the subsequent 12 launches were successful. The first launch of a recoverable satellite occurred in 1975. It weighed about 2 tons and operated in near-Earth orbit. Indeed not only was the LM-2 more reliable than the FB-1, it was able to lift heavier payloads. The two vehicles were so similar that for many years Western observers regarded them as the same. The FB-1 program, due to its difficulties and limitations, was canceled in 1981.

The 1974 failure of the LM-2A served a purpose in that subsequently extensive revisions were made to the wire harness and prelaunch checkout procedures to improve redundancy and reliability. These changes resulted in the designation of LM-2C being used for subsequent vehicles of this type. The LM-2C was the basis for Chinese commercial launch services (the first commercial activities involved piggyback payloads only), and provided the foundation for future developments. Again, as the missile/rocket technology is largely symbiotic, mention needs also be made of the CSS-4 ICBM program. Between January and November 1979 at least five CSS-4 test launches were made, furthering the technology base.

The Chinese desire to launch their own geostationary satellites led to the development of the LM-3 vehicle. Planning for a geostationary communications satellite program began in the mid-1970's, including not just the development of the satellite and the LM-3 launch vehicle, but "the establishment of the launch site in Xichang, the tracking, telemetry, and control (TTC) stations on Earth and compatible Earth communications service stations. Two geostationary communications satellites were launched, in 1984 and 1986."[27] See Figure 3.1. The basic change from the LM-2C to the LM-3 is the addition of a high-performance third stage. This is particularly important, however, as this third stage utilizes cryogenic propellants (liquid oxygen and hydrogen). Use of cryogenic fuel placed China as only the third country in the world capable of doing such, along with the United States and France. From its launch site near Xichang in Sichuan province, the LM-3 is capable of placing a 1400

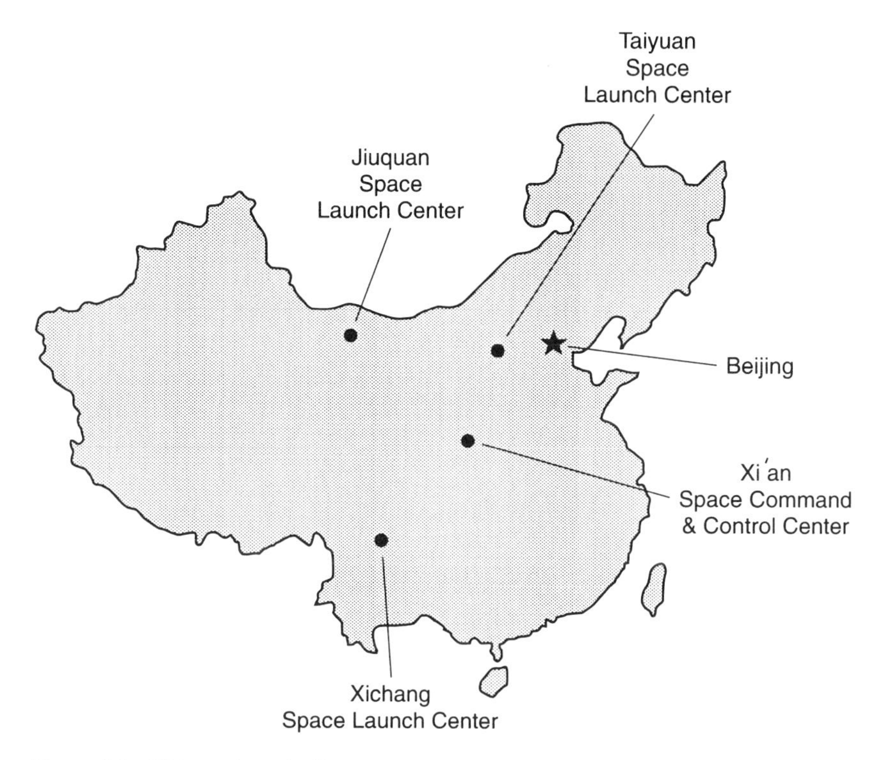

Figure 3.1 Chinese launch sites.

kg spacecraft into geostationary transfer orbit (GTO). The LM-3A is a "stretch" version (of the first two stages), with an enlarged third stage using two improved cryogenic engines.

One example of China being hindered by its own isolation can be found in the area of remote sensing. China was introduced to remote sensing in the mid-1970's and its potential utility in conjunction with some of the issues inherent to a country of the geographic expanse of China immediately became apparent. As remote sensing was a new field of space applications, however, and with the educational gap left by the Cultural Revolution, there were few in China with any exposure let alone expertise in the field.

Emphasis was immediately placed within China on developing skills for utilizing remote sensing as part of China's efforts toward national economic development. These have included resource surveying, environmental monitoring, national land management, and regional development. Between the mid-1970's and 1989, the National Remote Sensing Center (NRSC) trained more than 800 remote sensing specialists to work on various related projects.[28] Those projects have included: a land resources inventories at the national, regional, and subregional levels, including compilation of a 1:2,000,000 controlled land use satellite image map of the country using optical rectification techniques; production of over 20 thematic maps on soil, land use and land type, and flood history analysis for a 4000 km^2 in the area of the Huang-Hui-Hai plain; and both a geological map on the scale of 1:1,500,000 covering the Qinghai-Xizang plateau and a linear structure map of China on the scale of 1:6,000,000, compiled using Landsat data. Remote sensing techniques have also been increasingly utilized in disaster prevention, mitigation, and damage assessment.

Next in line within the Long March launcher family was the LM-4, consisting of a stretched LM-2C first and second stage topped with a new third state using storable propellants. It was originally designed in the late 1970's for Chinese geostationary communications satellites by the Eighth Academy (Shanghai) as part of the competition between the Eighth and First Academies. Lift capabilities proved less than sufficient, however, and it has been used exclusively to launch polar orbit meteorological satellites (Feng Yun series, FY, meaning "wind and cloud") which are also designed and built in Shanghai. The LM-4 is primarily launched from the site near Taiyuan in Shanxi province, which likely was originally used in the 1970's as a test site for the CSS-4.

In 1985, after serious consideration the Chinese government decided to make satellite communications a high priority program. A program called "Leasing for Transition" was initiated, where transponders were leased or purchased from Intelsat and other satellite owners in order to establish a Chinese operational domestic satellite system for communication and television broadcasting on an incremental basis. It was designed as a three-part program. Phase One relied on Intelsat transponders for pilot experiments and demonstrations, while the construction of ground networks and manufacture of equipment were underway. Phase Two utilized communications satellites made and launched by China, using leased and purchased transponders as a supplement. The third stage involved use of high-capacity communications satellites made and launched in China to form an operational domestic satellite communications satellite.

"Prior to 1985, there was only one nationwide television channel, the China Central Television (CCTV), Channel 1, transmitted through microwave links, with multistage relays and videotape (provincial television coverage relies mainly on the latter methods). Even then, the total populations with access were estimated at 62% by the Chinese, with reception quality uneven at best."[29]

Parallel development of improved launchers continued also. Utilizing existing hardware, the Chinese developed the LM-2E as a way to provide substantial lift capability to low Earth orbit at a modest cost. Its first test flight was in

July 1990. A proposed follow-on is purported to be part of the Chinese manned space effort as it would provide heavy-lift capability to low Earth orbit.

Concurrent with development of a full family of launchers, the Chinese also developed a full compliment of satellite technology. According to Chinese sources, China had successfully built and launched 37 satellites of different types through September 1996.[30] Among these indigenous satellites, the Chinese tend to group them into three basic categories: scientific exploratory technological experimental satellites; communication and broadcast satellites; and satellites for space physical exploration.[31]

The first category, the scientific exploratory technological experimental satellites, are the recoverable satellites. Each weighs about 2 tons, and is comprised of two modules. The upper module is the recovery module, the lower being an equipment module. Each has been launched into near-Earth orbit with an apogee of about 400–500 kilometers and a perigee of about 200 kilometers by an LM-2. After a scheduled number of days of operation in orbit, the recovery module is released and returned to a target area in China. Various payload tests can be performed on the satellite platform, including remote sensing, materials processing. As three-axis stabilization is used for the satellite in orbit, it has microgravity of 10^{-4} and 10^{-5}.

Chinese communication and broadcast satellites, the second category, are spin-stabilized geostationary satellites. Originally referred to by the acronym STW, for Shiyan (experimental) Tongxun (communications) Weixing (satellite), only the first two were considered experimental and subsequently have been called Dong Fang Hong (DFH)1 or 2. Cylindrical in shape with a diameter of 2.1 meters and a height of 3.1 meters, the first was launched in 1984, on the LM-3 which was specifically developed for that task. Another was launched in 1986. Through utilization of these two satellites, the intended system of communication and broadcast capabilities was declared successful after a trial operation period of 2 years. Very early on, the Chinese used their success in this area for marketing purposes. In a presentation in early 1987 in the United States, Chinese officials spoke about the potential of their satellites.

> Different payloads can be installed on the platform of the satellite. The power generated by the solar cells can supply eight to 10 transponders of four to six watts. The territory and district of a small or medium-sized country can be covered by such a satellite if different narrow-beam antennas are used. The power supply system and on-board fuel storage can keep the satellite working for as long as seven years. Since these satellites are technically reliable and low-cost, they are worth being considered for a range of practical applications.[32]

The DFH-2 only had four transponders and was eventually found not to be economically competitive.

The third category, satellites for space physical exploration, has been limited. SJ-2 (Shi Jian, or experimental) is one of the three satellites launched by a single launch vehicle in 1981.[33] SJ-2 weighed 257 kilograms, with four expandable solar panels on the top of the satellite. Spin stabilization was employed, with the attitude control system keeping the spin axis of the satellite pointed at the Sun to assure that the solar panels had sufficient power and that the Sun could be observed from the desired position. Instruments on board included: thermal ionization gauge, solar x-ray probe, solar violet-ray probe, magnetic field strength gauge, sun sensor, scintillation counter, far and near infrared radiometer, and electron direction detector.[34] Collected data was provided to scientific research organizations in China.

China's space program is reflective of many mirrors, past and present. Like Japan, China's first space launch occurred in 1970. Also like Japan, China has limited the number of launches it undertakes annually. Like the United States, China has both a military and a civil space program. In China, however, the programs bifurcate at the applications level, as opposed to in the United States where they are completely separate at the research and development levels. That means that whereas the United States in the post-Cold War era is engaging in an effort to converge at least some of the military and civil technology bases, that effort is not required in China. China is, however, following the lead of both the United States and Russia in undertaking conversion efforts within its space industry in an effort to raise capital. Unlike Russia, the undertaking has not been done by closing the government-funding faucet all at once, but rather more slowly to allow for a transition.

The existence of China's commercial space program is directly attributable to Deng Xiaoping's policies on economic reform. In March 1978, Deng Xiaoping called for the Chinese defense industry sectors (which includes space) to devote their efforts to economic growth. This actually occurred prior to the formal commencement of Deng's general economic reform program in December 1978. The second was a defense industry reform program initiated in early 1983, which called for the systematic transformation of defense-oriented industries into civilian-oriented industries. This latter conversion policy meant that all the state-owned space enterprises almost immediately began to see their budgets dramatically cut.

As in other fields, space organizations began to supplement their budgets through entrepreneurial activities. The conversion process was neither swift nor easy, however, with almost predictable consequences. Salaries among space engineers were said to be lower than that of some street vendors, with subsequent low morale and high personnel losses. Students, not surprisingly, increasingly began looking for work in the private sector rather than high-tech career paths dependent on government budgets for both schools and industrial support. The negative impact of personnel issues on the space industry spiraled further down when coupled with the impact of insufficient budgets on infrastructure and equipment maintenance and procurement.

In 1984, rumors within the U.S. national security community were strong that the Chinese had built a new launch site near Xichang. Further, considerable activity in that area, e.g., trucks, equipment, and people moving in and out, fueled speculation that a launch date was approaching. In January 1984, the Chinese did indeed make their first attempt to launch a geosynchronous satellite (STW1, an experimental communications satellite). The second stage did not fire, however, due to a thermal control problem. NORAD immediately picked up the launch and data on attitude, altitude, etc., was published in *Aviation Week & Space Technology.*[35] That they had launched from about latitude 28 degrees N made it evident that they had copied U.S. systems. When Sino-Soviet relations soured in 1956–57 and technology transfer was blocked, the only information the Chinese had was from open literature, which primarily discussed U.S. systems. Kennedy Space Center (KSC) is at about latitude 28.5 degrees N. Hence, the Chinese launch site position was not a coincidence; rather they had picked a similar location to make it easier to emulate procedures and expectations that they had read about.

Reports of the attempted Chinese launch published in the U.S. press reached the Ministry of Space Industry (MASI), the Chinese organization which at that time was in charge of space.[36] Since secrecy was now a moot issue anyway, the case was made that the time was propitious to open the Chinese space program to the world, as the potential for revenue generation from launches was correctly viewed as substantial. Once that policy decision was made, it became clear that China would need to try to establish a reputation for openness and honesty, and a favorable reliability rate with their launchers, in order to compete with the European Ariane launcher.

After several months of consideration, the Chinese began to build a plan for commercialization of their launch industry. They knew they needed (1) a budget, (2) advertising and marketing plans, (3) a niche, and (4) pricing at about 80–85% of the current market price. They knew they also had to made the launch site more user friendly, in terms of a hotel, airport, and communications facilities. The China Great Wall Industry Company (CGWIC) was already in existence and had an office in Washington, D.C., although previously it had served as little more than a front for MASI. In 1985, it officially became responsible for marketing commercial launches.

Bureaucratic politics was important during this shaping and posturing period, as it still is today. For example, in 1984 the Chinese began looking at a direct broadcasting system (DBS). Deng authorized an allocation of $300 million to initiate work in that field. The plan got approval from the Ministry of Film, Radio and Television (MFRT). The Ministry of Post and Telecommunication (MPT) did not approve, however. Specifically they criticized the Ku-band approach, saying that with modification the C-band could also be included. At this point, Minister Lu at MFRT made a tactical error. The Long March technology was still new, and he rejected it as a launcher candidate for DBS. Rejecting Long March alienated CASC, in charge of the Long March, which joined forces with MPT. Subsequently, the project was canceled at the last minute.

After about a 1 year preparation period, the Chinese made their first sales trip to the United States in the spring of 1985. Their first sales pitch was made to COMSAT. This coincided with a generally increased presence of Chinese space officials at international conferences, as both observers and participants. There were no large, umbrella space agreements between the United States and China, agreements were negotiated on a case-by-case basis.

Although the LM series was made available on the international market in 1985, no firm launch orders were received for some time. "Only after China was able to demonstrate technological competence and ensure the safety of payloads did the international market begin to take China's product more seriously."[37] The subsequent decisions by both Aussat and the AsiaSat consortium to launch spacecraft from China resulted in considerable opposition from Western launch suppliers concerned about their own market share, and caused concern in the U.S. defense community.

The United States at that time maintained a very tight control over the transfer of technology which is perceived to be of a sensitive nature or of strategic value to other countries through the International Traffic in Arms Regulations (ITAR) administered by the Department of State (DOS). Under the ITAR there were restraints on the transfer of all manner of hardware and data to other countries of items of the "Munitions Control" (MC) list, which includes communications satellites and related equipment.

> The prospect of the Aussat and AsiaSat missions being launched in China drove the U.S. government in December 1988 to conclude the Memorandum of Agreement on Satellite Technology Safeguards Between the Government of the United States of America and the People's Republic of China in order to preclude the unauthorized transfer of sensitive technology associated with the possible launching of US-manufactured Asiasat and Aussat satellites from the People's Republic of China. This agreement essentially confined the transfer of technical data to that required by the Chinese to perform their duties as a launch services provider. It also detailed constraints on access by Chinese citizens to the spacecraft processing facilities at the Xichang launch site as well as the transport of the spacecraft within China and the procedures to be followed in repatriating spacecraft debris in the case of a launch failure.[38]

Although the agreement was written specifically to cover the Aussat and AsiaSat programs, the points made within it were considered generic to any launch in China under the control of the United States. Because Western spacecraft initially used a significant number of U.S.-sourced units, components or materials, the United States effectively controls export issues for any Western spacecraft and the 1988 agreement immediately became relevant to all Western satellite users. This points out two critical points: that the Chinese market is such that international businesses are eager to get in to the extent of perhaps recklessness, and that U.S. businesses are put in a particularly difficult position. However, a "Technology Control Plan" stating how technology transfer control would be implemented was also an integral part of the launch service contract with the Chinese.

Also, other launch vendors were concerned about China from the perspective of China being able to seriously undercut any other prices because of inexpensive Chinese labor costs and negotiated prices within China. A second Memorandum of Agreement was hence concluded in late 1988, under threat of trade sanctions. The thrust of this agreement is the imposition of "market economies" on Chinese launch service pricing as well as limiting the number of launches before the end of 1994 to 11. The following extract gives further details of the agreement.

- The Agreement supports the application of market principles to international competition in commercial launch services.
- The PRC will take steps to ensure that its providers do not materially impair the smooth and effective functioning of the international market for commercial launch services. Among these steps, the PRC will ensure that any support to its providers is in line with practices of market economies.
- The PRC shall also require its providers to offer their services, including insurance and reflight guarantees, at prices, terms and conditions on a par with those offered by providers from market economies.
- The PRC will prevent its providers from offering introductory or promotional prices except for the first or, in extraordinary cases, second successful commercial launch of a new vehicle.
- PRC providers shall not launch more than nine communications satellites (including the two Aussat and one AsiaSat satellites) for international customers during the agreement, which shall terminate December 31, 1994, unless extended.
- PRC commitments to provide commerical launch services are to be proportionally distributed during the agreement.
- The PRC will not offer inducements of any kind to international customers which would create discrimination against providers of other nations, the PRC will also require that its providers not discriminate unfairly against international customers or suppliers.

- The PRC and the US will consult on the agreement and related matters annually and within thirty days upon request.
- During annual consultations, the limit on the number of PRC launches may be reconsidered upon request by the PRC in light of unforeseen developments. The US will decide on such a request within thirty days.
- The US and the PRC are prepared to discuss comprehensive multilateral rules for commercial launch services.
- A comprehensive review of the agreement will begin in September 1991.
- Any application for a US export license will be reviewed on a case-by-case basis. Nothing in the agreement shall be construed to constrain the US from taking any appropriate action consistent with US laws and regulations.[39]

The more recent Sino-U.S. agreement signed in January 1995 provided for up to 11 additional geostationary launch contracts through the year 2001. The agreement requires China to price its commercial launches within 15% of what Western launchers would charge for comparable services. The agreement further limits China to launching 15 Western satellites into geostationary orbit through 2001, which includes 4 carried over from the previous agreement. "However, if during the first three years that the accord is in effect the average number of launches worldwide is 20 or more, the ceiling would be raised to 17. If that 20-launch trend continues for a fourth year the ceiling would be raised to 20."[40] It is interesting that low Earth orbit launches, which are planned for such systems as Iridium and Globalstar, were not included in the agreement.

Conclusions

The turbulent years within which the Chinese space program developed ought to indicate the tenacity of the people and the willingness and ability of the government to act independently of the outside world if necessary. The results of those independent actions, however, do not necessarily coincide with either the plans or desires of other countries, individually or in combination. Although nonverbal communication can be influential, it is certainly not the desirable route over the long term.

China has developed its space program to meet its strategic goals, commercial goals, and domestic goals. Likely commercial goals have priority now, in order to meet the domestic goals which currently shape Beijing's most pressing strategic goal, internal stability. Unless maintaining Chinese internal stability runs contrary to U.S. policy, perhaps the United States ought to pay close attention to Beijing's plans for the future and see where the United States might become involved, and hence influential.

Endnotes

1. Zhang Xinzhai, "The Achievements and the Future of the Development of China's Space Technology," *Aerospace China*, Summer 1996, 22.
2. Walter McDougall, "*. . . the Heavens and the Earth*," (New York: Basic Books, 1985) 177–194.
3. John Wilson Lewis and Xue Litai, *China Builds the Bomb*, (Stanford, CA: Stanford University Press, 1988) 52.
4. For a comprehensive history of missile/rocket development in China, see: John Wilson Lewis and Hua Di, "China's Ballistic Missile Programs," *International Security*, Fall 1992 (Vol 17, No. 2) 5–40.
5. For a biography on Qian Xuesen, see: Iris Chang, *Thread of the Silkworm* (New York: Basic Books, 1995).
6. Perhaps because of its long history, few really "new" organizations are ever formed in China, rather there are mergers and reorganizations of old ones. In this case, the Scientific Planning Commission and the State Technological Commission were merged.
7. See: Gordon Pike, "Chinese launch services: A user's guide," *Space Policy*, May 1991, 103–115. This program was later renamed the DF-1 after the original DF-1, renamed the DF-3, was abandoned in 1963.
8. Chang, 1995, 214.
9. *China Today: Defense Science and Technology*, Vol 1. (Beijing: National Defence Industry Press, 1993) 356.
10. *China Today*, Vol. 1, 356.
11. Lewis and Xue, 138.
12. Chang, 216–217.
13. Chang, 217.
14. Chang, 217.
15. Lewis and Hua, 14.
16. Chang, 222.
17. The Chinese sometimes use the designation CZ, which stands for Chang Zheng, which is Chinese for Long March.
18. Chang, 222.
19. Lewis and Hua, 17.
20. Fisher, 41.
21. The Fifth Academy had been renamed the Seventh Ministry of Machine Building in 1965.
22. Chang, 249–251.
23. Named for the melody it was broadcasting, The East is Red. Zhang Xinzhai, 1996, 22.
24. Wu Guoxiang, "China's space communication goals," *Space Policy*, February 1988, 42.
25. Pike, 106.
26. Pike, 106.
27. Liu Ji-yuan and Min Gui-rong, "The progress of astronautics in China," *Space Policy*, May 1987, 142.
28. Changchui, 66.
29. Wu Guoxiang, "China's space communication goals," *Space Policy*, February 1988, 42.
30. Zhu Yilin and Xi Fuxiang, "Status and prospects of China's space programme," *Space Policy*, February 1997, 70. The number of satellites launched can differ according to

sources because of the inclusion or exclusion of some payloads on various lists. See, for example, *Jane's Space Directory, 1996–97*, ed. Andrew Wilson, Jane's Information Group, Inc. Alexandria, VA, 195, which gives a slightly different list of satellites launched by China than that of Zhu and Xu.

31. Liu and Min, 143–144.
32. Liu and Min, 143.
33. It is noteworthy for both commercial and strategic reasons, however, that the Chinese are able to launch multiple satellites from one launcher.
34. Liu and Min, 143.
35. "Chinese Launch," 6 February 1984, 25.
36. MASI evolved from the Fifth Academy and then the Seventh Ministry.
37. Jian, 23.
38. Pike, 111.
39. "Fact Sheet on the Memorandum of Agreement Between the US and PRC Regarding International Trade in Commercial Launch Services," U.S. Department of State, Washington, DC, 1989.
40. Warren Ferster, "U.S. Says China Violated Accord," *Space News*, 19–25 May 1997, 18.

Chapter 4

Chinese Space Players

Introduction

Part of the *Wei Qi* aspect of the Chinese space program is that finding one's way through the byzantine Chinese bureaucracy can be a mind-boggling feat. Starting from the assumption that bureaucracies are inherently complex, the Chinese have had some 5000 years to complicate theirs beyond the norm. Further, by their own admission, they are excruciatingly over-bureaucratized. China has 864 National Science and Technology Institutes. Although a 1995 Chinese government report called for turning most research laboratories into business centers dedicated to spurring economic growth,[1] the problem with doing that goes back to maintaining a proper balance between the need to disband state owned enterprises and the political need to maintain employment levels. Combining the sheer number of organizations involved with the multiple restructurings which have occurred in Chinese space organizations to reflect the new commercial bent of operations, the latest in 1993, can create explanations for "Who's in Charge" similar to the dialogue of the Abbott and Costello "Who's on First?" routine. There are also countless organizations within organizations. Further, references are still made to organizations by their former names, which exacerbates confusion and results in differing references being made to the same or officially nonexistent organizations.

For example, in a 1996 Western publication it was stated that "The Shanghai Bureau of Astronautics, a research and production base under the Ministry of Aerospace Industry [MASI]. . . . ."[2] yet MASI officially disappeared from the organization charts after the 1993 reorganization, and the Shanghai Bureau of Astronautics has been renamed the Shanghai Academy of Spaceflight Technology (SAST).[3] It is not just Westerners that make such references either. The *Beijing Review* stated in 1997: "The Ministry of Astronautics Industry proposed 3,972 research project. . . . ".[4] In many if not most cases, the offices, personnel, and responsibilities remain the same within the organizations, so making sure the official name is correct may or may not be relevant. From an organizational security perspective, the Chinese are likely delighted by this continued confusion. Although a colleague's suggestion that perhaps this was a cunning bit of Chinese infowar likely gives the Chinese more credit for deliberate strategizing than is warranted, the effect is nevertheless similar. That being the case though, I will nevertheless attempt to provide a brief overview of the key organizational players in the Chinese government and the space program, including lineage as best it can be determined.

Being mentioned with little detail here usually indicates that the organization was mentioned or cited in interviews or literature at least once as "important," with little subsequent mention or reference. This I have taken to mean that it is an organization within another organization, and is likely influential internally or within a specific field. In some cases, when possible in interviews, I did ask for more specifics about an organization and was simply ignored or the reference was suddenly "forgotten." The "why" behind this problem is important.

As in the Soviet Union, and now Russia, not just the technology of civil and military space activity is symbiotic, but the institutions guiding and developing it are as well. Although this can be beneficial in terms of maximizing technology research and development funds and efforts, it can also be detrimental. The Chinese today seem almost diametrically pulled between the necessity of opening their space program to other countries for economic development purposes, and paranoia about revealing anything other government officials may deem national security related. Hence in a time when other countries are recognizing the need to "know your partner" in civil program cooperation, the barriers are still well intact in China.

Also like the former Soviets, the Chinese utilize a parallel government system whereby at each level (central, province, etc.) there are both government mechanisms and party mechanisms, with the party holding ultimate power. Unlike the Soviets, however, the Chinese also made the military a third partner. That means that the People's Liberation Army (PLA), which includes all the services, can be used not only for international security purposes but domestic policy goals as well, and is in many respects beyond the government's jurisdiction. The PLA holds a bureaucratic rank equal to that of the State Council, which is the government's highest level of authority.

Within these three vertical organizational pillars of China's government, there are also multiple connecting horizontal bars. For example, as the party is still the holder of ultimate power, another aspect of bureaucratic organization that must be considered is the *xitong*. *Xitong* literally means "system." They are functional groupings of bureaucracies that together deal with a broad task the top political leaders want performed, usually headed by a powerful party member. The various *xitong* are the party arm of bureaucratic organizations, adding yet another circle to the dizzying web of interrelationships within which decision making takes place in China. *Xitong* cover a wide variety of areas, though six have been identified as particularly relevant to the management of the country: Party Affairs, Organization and Personnel, Propaganda and Education, Political and Legal Affairs, Finance and Economics, and the Military. Once decisions are made at this level, they can then be safely passed to the government for implementation.

One might argue that the Chinese system is organizationally confusing but in practice operates much like in the West. Certainly, stovepiped organizations, the interagency process and separation of powers (and confusion) can be found in the West and could be analogized to various Chinese counterparts. Also, governmental agreements with subsequent awarding of private contracts and subcontracts are common. But the analogy does not hold in a number of senses: as referenced earlier as critical, the legalities restraining institutions in the West are missing; the open scrutiny of decision making is missing; dissemination of knowledge amongst parties is severely limited; and the behind-the-scenes "Big Brother" presence of the party overshadows everything.

The Maze

The State Council

Established in 1954 to replace the transitional Government Administrative Council, the State Council is headed by the premier and has 14 members. It includes commissions and ministries within its purview, with commissions responsible for issues which can involve multiple ministries. The State Planning Commission, for example, is in charge of long-term and annual plans, and the State Economic Commission is responsible for resolving interministerial issues resulting from plan implementation. Much of the work of the ministries is focused on development and the urban economy, while in the rural regions the party is still dominant. The commission directly relevant to space is the Commission of Science, Technology and Industry for National Defense (COSTIND).

Space Leading Group (SLG) in the State Council

Founded in April 1989, this group was described in Yanping Chen's 1993 article[5] as the top group responsible for policy making and mission coordination among the central government agencies. Its purpose was stated as coordinating coherent space activities in China, similar to an organization called the Central Special Committee that existed in the 1960's and 1970's led by Zhou Enlai. That committee was, like many others, disbanded during the Cultural Revolution.

The members of the SLG, as of 1993, were stated as being "the Prime Minister of the State Council, the Chairman of COSTIND, the Vice-Chairman of the State Committee of Science and Technology, the Minister of Aerospace Industry, the Vice-Minister at the Ministry of Foreign Affairs, and the Vice Chairman of the State Committee of Central Planning."[6]

The role of the SLG today is unclear. As stated previously, although several people in positions which ought to have rendered them full knowledge of such a key organization were specifically asked about it, nobody seemed to know or be willing to acknowledge it. Others summarily dismissed it as now defunct.

The Commission of Science, Technology and Industry for National Defense (COSTIND)

The National Defense Science and Technology Commission (NDSTIC), discussed earlier as the original oversight organization for both missile and nuclear weapons development chaired by Nie Rongzhen, and the National Defense Industry Office were merged in 1982 to form COSTIND. It is directly tied to the State Council and is responsible for both aerospace and strategic weapons, on the civil side controlling the launch sites and tracking, telemetry, and control (TT&C). It is also the coordinating organization for space, and as such divides both projects and resources among interested parties, e.g., the Ministry of Film, Radio and Television (MFRT) and the Ministry of Post and Telecommunications. Hierarchically, the China Aerospace Corporation (CASC or CAC) and the China National Space Administration (CNSA) are under COSTIND. Project consideration begins in COSTIND for CASC. Although COSTIND theoretically has no direct authority over CASC, its influence and control are significant, through several mechanisms. First, COSTIND is CASC's link to the State Council, which is the ultimate authority for funding. Second, and more importantly on a day-to-day basis, COSTIND's power is via control of personnel. COSTIND appoints the chief and deputy chief designers for key projects within CASC's purview.

Recently, COSTIND leadership transitioned from Ding Henggao to Cao Gangchuan. In a somewhat surprising move, Cao was appointed from outside. The deputy director was and remains Lt. Gen. Shen Rongjun. He is the senior military officer in charge of China's space efforts. In an interesting example of the Confucian-based family tie system which is prevalent in China, Ding is married to Nie Li, daughter of COSTIND founder Nie Rongzhen. Apparently concern that family influence was becoming too strong even by Chinese standards, and a bad property deal, led to Ding's downfall.

State Science and Technology Commission

This group, headed by Song Jian, is responsible for developing macro-level policy concerning space, primarily involving research and academics.

Chinese Aerospace Corporation (CASC or CAC) and Chinese National Space Administration (CNSA)

In 1956, the party's Central Military Commission created the Fifth Academy within the Defense Ministry to be in charge of missile research and development. Within that organizational scheme, a number of sub-academies were also created, for example, the First Sub-Academy was in charge of the general configuration and rocket engines and the Second Sub-Academy was in charge of guidance systems. Many of these were subsequently renamed. The First Sub-Academy became the First Academy in January 1965, generally responsible for carrier rocket research. In 1965, the Fifth Academy became the Seventh Ministry of Machine Building. Then, in 1982, that organization was redesignated as the Ministry of Space Industry (MASI). There were multiple academies within the MASI purview, many of which sold their products overseas after the Chinese push for capital in the 1980's. There was, for example: the First Academy, the pedigree of which was previously mentioned; Second Academy (producing antiaircraft missiles for sale abroad); the Third Academy (producing antiship missiles for sale abroad); and the Fourth Academy, in charge of solid rocket motors. Companies were then specifically set up as outlets for these sales. China Precision Import-Export Company (CPMIEC) was an export company set up for missile and arms trade, for example. Also, as was brought out already, competition even surfaced within the academies. Those academies which were successful in export sales were rewarded twofold: by showing their relevance to the national economy they were rewarded with government funding and they were also able to supplement their government funding with part of their "earnings." Indeed the director of First Academy is said to have told his colleagues: "For money, develop the DF-15 (missile); for fame, develop the CZ-3."[7] This pull of forces, and subsequent rewards, continues today. Indeed the DF-15 was developed for hard currency by the First Academy, and it likely could bring in $200,000 on the arms market.[8]

Until 1984/85 all of these were classified organizations along the lines of the National Reconnaissance Office (NRO) in the United States and very much military dominated. Even by Western standards, MASI was a large fully state-owned entity. It employed about 55,000 technical personnel and a support workforce of about 250,000, and operated approximately 100 factories and 80 research and design institutes.[9] Today CASC and CNSA have assumed the responsibilities (personnel, buildings, etc.) of MASI.

The principal role of CNSA is to serve as China's policy organization and interface with other national space agencies, while CASC exerts primary control over the national space program on a day-to-day basis. Basically, CNSA handles external matters while CASC handles internal matters. There is considerable sharing of personnel between CASC and CNSA. For example Hua

Chongzhi and Lao Ge are both deputy directors for the Department of Foreign Affairs of both organizations.

Today, CASC is a corporation with 270,000 employees, including over 100,000 engineers. CASC primarily engages in research, design, test, manufacture, and commercialization of various space technological products and civilian applications. Further, as a national level company, launch vehicles, satellites, and other space products in China are within its exclusive domain. CASC has achieved success in the fields of launch vehicle technology, cryogenic propellant rocket technology, satellite retrieval, and multipayload launches, and possesses the capability of launching satellites into low Earth orbit (LEO), Sun-synchronous orbit (SSO), and geostationary Earth orbit (GEO).

China and CASC are particularly proud that space spin-offs have been widely applied to multiple sectors of the Chinese national economic development effort. Beyond satellite applications and ground equipment, the main civilian products include electronics, automobiles, communications, computers, automatic control systems, petrochemical equipment, medical apparatus, packaging machinery, and consumer electronics.

To date CASC has established technical cooperation and trade relations with more than 50 countries, regions, and international organizations. Almost 100 joint-ventures enterprises are set up in China and abroad.

There are multiple institutes under CASC (see Figure 4.1), and much like the former "bureaus" in the Soviet Union, some have been able to respond to commercialization efforts more than others.

China Academy of Space Technology (CAST)

Formerly part of the Fifth Academy, CAST was formed in 1968 specifically to be responsible for the design and manufacture of DFH communications satellites, FSW recoverable satellites, and Earth resources/remote sensing satellites. Under its purview, the 504 Institute in Xi'An, discussed earlier in conjunction with its Cultural Revolution perils, is responsible for commercial payloads.

China Satellite Launch and Tracking Control General (CLTC)

This organization was specifically established to provide commercial launch and tracking, telemetry, and control (TT&C) services. Like other Chinese space organizations, it is subordinate to COSTIND, but not through CASC as are most others. CLTC manages one aerospace command and control center (Xi'an, XSCC), three satellite launch centers (Jiuquan, Xichang, Taiyuan), one comprehensive TT&C network (Xi'an) and two research institutes. See Figure 3.1. In all, the organization claims a workforce of more than 20,000, of whom some 5,000 are engineers. Pictures of personnel at the controls during

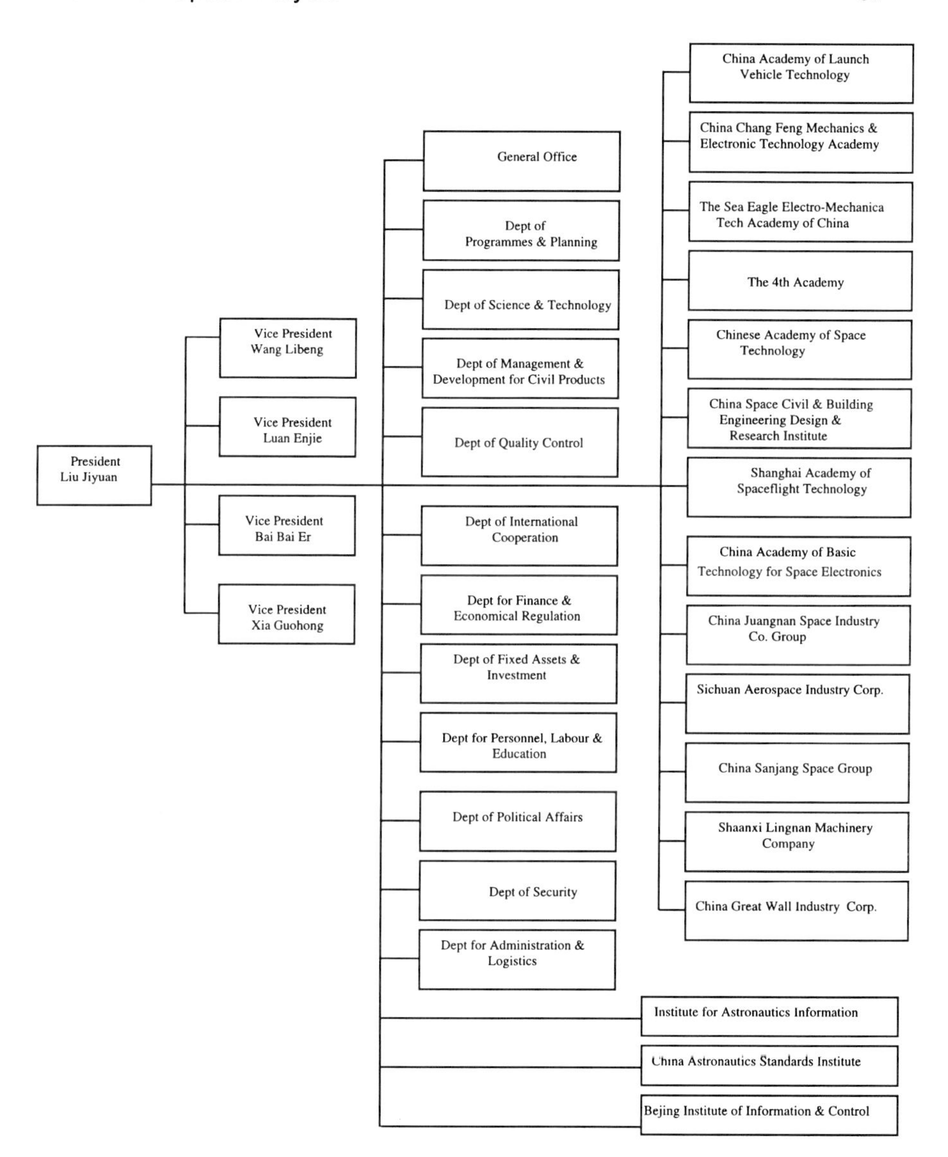

Figure 4.1 Organizational chart for China Aerospace Corporation (CASC) showing various bureaus and academies under its direction. *Used with permission of China Great Wall Industry Corporation.*

launches frequently show men in uniform.[10] According to a former TT&C personnel manager in China, the military still operates the launch sites. Spacecraft built by civilian organizations in China or abroad are handed over to the military security organization at the launch site. As such, the launch sites are ultimately owned by COSTIND, which then takes approximately one-third of the profits from commercial launches to cover fuel, infrastructure, and personnel costs. Combining the facts that the launch sites are under the purview of the military and the military is a key jobs program within China, and that aerospace generally is among the fields from which the government is reluctant to disband employees, few personnel cutbacks are likely and inefficiencies are said to abound. Indeed one Chinese launch expert speculated that as many as one-fifth of the launch personnel could easily be eliminated, but without much optimism for improvement in the near future.

Located in the ancient city of Xi'an, XSCC manages China's spacecraft TT&C network. This includes the command and control center, fixed and mobile stations, instrumentation ships, and reentry instrumentation airplanes. XSCC has served its role since the launch of China's first satellite in April 1970.

Jiuquan Space Launch Center (JSLC) is located in northwestern China at the Gobi desert. It is the earliest and largest satellite launch center in China, created in 1958 for military launch activities. It is primarily used to launch to medium and low orbit inclinations, including recoverable capsule launches.

In the 1960's and 1970's Mao decided to spread out the launch facilities for strategic reasons, primarily in response to the Soviet threat. Sites were specifically selected for their remoteness in accordance with his Third Line (*San Xian*) strategy, and built for military purposes. Other considerations for sites came into play also. That the latitude coordinates of Xichang are similar to those of Kennedy Space Center (KSC) is thought likely not coincidental. As mentioned earlier, after the Sino-Soviet split, the Chinese still had considerable Soviet space hardware to work from and the biggest actual loss was in utilization knowledge. For that they had to rely on published data, which came primarily from the United States. Launching from the same latitude as KSC allowed the Chinese to emulate the expectations for such technical points as the proper rocket attitude and altitude.

When the Chinese decided to enter the commercial launch field in the 1980's, they sought technology from the West for improving capabilities, including sending delegations to the United States and Europe. Because of the obvious dual-use technology transfer considerations, however, not surprisingly no other countries would or could sell to them. The Chinese have subsequently concentrated on upgrading their facilities indigenously, and spent considerable funds on such, sometimes without badly desired and needed near-term payback. Indeed facilities for launching LM-2E rockets from Jiuquan were assembled for a potential Globalstar contract, which then went elsewhere.[11] With pragmatic assessments for returns currently dominating Chinese decision making regarding funding, these are difficult investments to justify based only on potential, but difficult to refuse based on potential also.

Taiyuan Space Launch Center has been used for polar orbits since 1988.

Xichang Space Launch Center (XSLC) is located in Sichuan province, in southwestern China (Figures 4.2, 4.3, 4.4). It is located at latitude 28 degrees N and an altitude of 1500 m. XSLC is comprised of two launchpads, one dedicated to the LM-3, and the other used for the LM-2E, LM-3A, LM-3B, and the new LM-3C, and mainly used to launch geostationary spacecraft. All domestic communications satellites and commercial satellites are launched from

Figure 4.2 Launch pad at Xichang for Long March 2E/3A/3B launchers. *Courtesy of Rob Preissinger.*

Figure 4.3 The older of the two launch pads at Xichang, it is used exclusively for the Long March 3. *Courtesy of William Pendley.*

there. The second pad, for the LM-2E, was built in 14 months to better position China to respond to the commercial market. Again, an investment was made from which a return is expected and needed.

China Academy of Launch Vehicle Technology (CALT)

The former First Academy, which was also known at one time as the Beijing Wanyuan Industry Corporation, this organization was founded at the outset of the Chinese missile technology program in 1957. CALT and the Shanghai Academy of Spaceflight Technology are primarily responsible for launch vehicle construction, competitively on most and jointly, as in the case (solely) of the LM-3. CALT managed the development of the LM-2C and hypergolic and cryogenic engines.

Figure 4.4 The approach to the Xichang launch facility, showing the LM-3E pad on the left and the LM-3 pad on the right. *Courtesy of Rob Preissinger.*

Shanghai Academy of Spaceflight Technology (SAST)

Founded in 1969 and formerly known as Shanghai Bureau of Astronautics, SAST is now responsible for the first and second stage structures on the Long March-3 launch vehicle and is responsible for the Long March-2D and Long March-4 boosters. SAST is also known to have developed the Feng Yun-2, a geostationary meteorological satellite,[12] as well as a medium-range surface-to-air missile for the export market, capable of tracking multiple targets,[13] evidencing its proficiency in both civil and military technologies.

Even during this period of privatization, these bureaus and academies are still dependent on government funds because of the dual-use nature of their work. There has, however, been limited competition initiated between the formerly exclusively specialized institutions. For example, the Sinosat telecommunications satellite being built as a joint venture between China and the German aerospace firm DASA was opened to competition within China by CASC. CAST was institutionally stronger in the field, with far more experience. SAST also bid for it though, based on the meteorological satellite work it had done. SAST planned to initiate an entirely new design, whereas CAST intended to use/modify the standard DFH-3 platform. A senior experts group was convened, and selected SAST. Two years later, however, Liu Jiyuan, president of CASC, intervened and gave the contract back to CAST, apparently for varied political reasons. CAST was said to have had sufficient satellite work at the time of the contract award to SAST to keep busy with other projects, but two years later found itself less busy than expected and began lobbying Liu for Sinosat. There was also a internal push to support the DFH platform, as well as out-and-out turf protection. Whether by intervention or interference, the proper word debated between two Chinese scientists as this story was being told to me, SAST found the contract it was awarded on merit suddenly taken elsewhere.

China Great Wall Industry Corporation (CGWIC)

Originally set up by MASI as a kind of legitimate "front" for the defense-related industries wanting to get involved in foreign sales, the CGWIC is now the exclusive organization in China, with legal status, responsible for launch service marketing, commercial negotiation, contract execution, and performance. CGWIC is headed by Zhang Xin Xia, replacing Zhang Tong, who died in a March 1997 drowning accident while visiting the European launch facilities in Kourou, French Guiana. Xin Xia was formerly the chief economist with China Aerospace Corporation, substantiating the importance being placed on economics within the Chinese space program.

Organizationally, CGWIC has "partnerships" with many if not most of the other Chinese space entities, to provide one-stop shopping for international customers. The uniqueness of this arrangement is important.

> CGWIC has two attributes that are different from other international space trading companies in the West. First, it is the only export window for Chinese space products in the international market. Second, it is the only legal entity pricing Chinese space products. The prices of space products reflect the negotiated costs from its Chinese partners and the profits of CGWIC.[14]

This quote reflects the difficulty working within the Chinese system discussed prior, whereby a magic price spontaneously appears to the customer through an intermediary, here the intermediary being CGWIC. Clearly, CGWIC is not a policy decision-making body, but rather an administrative body.

Second Artillery

The Second Artillery of the PLA is the Strategic Rocket Force. The intertwined nature of this organization with the others is significant. The First Academy (now CALT), for example, builds the carrier rockets utilized by the Second Artillery and hence is technologically tied to all other aspects of the Chinese space program. Historically, the Second Artillery has been strictly a tool of the higher ups, rather than an initiator of policy or strategy. Lewis and Hua say, for example: "The soldiers of the Second Artillery and their comrades in the First Academy merely imagined that nuclear strategy was a matter to be debated and decided upon by the leaders in the Central Military Commission."[15] Politically the Second Artillery is, however, likely the strongest component of the PLA, receiving far more modernization funds than some counterparts.

Ministry of Post and Telecommunication (MPT)

The MPT is a functional department under the State Council governing the post and telecommunications industry. It is responsible for the macro-control of China's communications industry, making overall plans, coordinating projects, and supervising operations.[16] There have been a multitude of so-called "private companies" created, which are actually under the purview of MPT, as providers of telephone services. China Telecommunications Broadcast Satellite Company (China Telecom) is by far the largest. Within approximately the past 5 years, however, other companies, such as China United Communications (China Unicom), have also emerged, as competition under the same master. The result for the consumer has mixed. Although China Telecom was forced to drop its prices as a result of the competition, Unicom was evidently not only quite disorganized in its business transactions, but somewhat arrogant in its conduct as well.

Chinese Academy of Sciences

This organization is most often referenced in conjunction with space science work being conducted by universities. Likely it plays primarily a coordinating role.

China Satellite Launch Agents of Hong Kong, Ltd.

This group was established to promote the commercial use of Chinese recoverable spacecraft.

Chinese Society of Astronautics

A technical organization which also is said to assist in overseeing space development.[17] Its role in this area, however, is quite nebulous.

People's Insurance Company of China (PICC)

PICC is a state-owned company which, *inter alia,* offers insurance to customers of the various Chinese launch service.

Conclusions

Navigating bureaucratic and organizational waters in any country, including one's own, can be a difficult feat at best. I have heard colleagues from Canada to India to the United States complain that finding out the proper office to contact for a specific piece of information can be as difficult a feat as discovering Unobtainium in the laboratory. In China, the difficulty is increased almost exponentially, as a function of the increased size of the bureaucracy, the compartmentalization of knowledge, and the always prevalent paranoia about divulging information to foreigners who will use it to exploit China.

The many organizational players involved with Chinese space development make it difficult if not impossible for many outsiders to peer into the inner workings of the Chinese system with any degree of clarity. Contacts made between foreign scientists and business contacts nurtured over long periods of time with the Chinese likely become the most fruitful. There can be an important difference, however, between these kinds of contacts and *Guan Xi*. Scientists in various disciplines, bound together by quests for knowledge presented by nature rather than government, tend to want to work together toward common goals, and subsequently share an incredible amount of information amongst themselves. Businesses can thrive or wither according to their contacts, making the sharing of knowledge within a company, especially one which rotates personnel as most in the Far East do, imperative. *Guan Xi*, in contrast, holds knowledge at an individual level, which perpetuates secretiveness and subsequently inhibits confidence building and trust through knowledge.

Hopefully, the Chinese will soon accept that organizational secrecy or secrecy through deliberate confusion does not support their goal of increasing international cooperation. That will be a first step in overcoming what is a natural boundary in any case of one country getting to know another. Allowing those who have the knowledge, and providing it to those in liaison positions likely to be asked, to answer a direct question can go a long way in building confidence among partners. That will then move China to the same place as the United States, Canada, India, or Russia, for example, where bureaucracies are simply annoying, rather than impossible.

Endnotes

1. Arthur Fisher, "A Long Haul for Chinese Science," *Popular Science*, August 1996, 37–39.
2. Mark Williamson, "Chinese Space Show," *SPACE and Communications,* November-December 1996, 29.
3. Nicholas L. Johnson and David M. Rodvold, *Europe and Asia in Space, 1993–94*, Prepared for USAF Phillips Laboratory, Kirtland AFB, NM 87117, by Kaman Sciences Corporation, Colorado Springs, CO, 15–16.
4. January 6–12, 1997, 15.
5. Yanping Chen, "China's Space Commercialization Effort, Organization, Policies and Strategies," *Space Policy,* February 1993, 45–53.
6. Chen, 1993, 48.
7. Lewis and Hua, 1992, 35.
8. Michael A. Dornhein, "DF-15 Sophisticated, Hard to Intercept," *Aviation Week & Space Technology*, 18 March 1997, 23.

9. Yangping Chen, "China's space commercialization effort," *Space Policy,* February 1993, 46. For further information on specific bureaus, see Nicholas L. Johnson and David M. Rodvold, *Europe and Asia in Space, 1993–94,* Prepared for USAF Phillips Laboratory, Kirtland AFB, NM 87117, by Kaman Sciences Corporation, Colorado Springs, CO, 15–16.
10. See: Mark Ward, "Exploding China's dreams," *New Scientist,* 16 March 1997, 14.
11. Although it is sometimes professed that these facilities were specifically built for the LM-2E as part of the Chinese manned space effort. See: Stefan Barensky, "China Unveils Heavy Launcher, Manned Spacecraft," *International Space Industry Report,* 23 April 1998, 1.
12. Michael Mecham, "China Plans Seven Missions for Long March Booster in 1997," *Aviation Week & Space Technology,* 11 November 1996, 25.
13. Michael Mecham, "China Displays Export Air Defense Missile," *Aviation Week & Space Technology,* 2 December 1996, 61.
14. Chen, 47.
15. Lewis and Hua, 20.
16. *The APT Yearbook,* 1997, 307.
17. "Chinese Detail Small-Satellite Efforts," *Aviation Week & Space Technology,* 14 October 1996, 33.

Chapter 5
Current Program

Introduction

It is impossible to provide full descriptions of everything China is currently engaged in concerning space. Indeed even cataloging what programs are where on a spectrum running from propaganda, to discussed, approved, being designed, built, and operational would likely be impossible. Luckily, it is also not necessary for the analytical purposes of this book. What is important, however, is to provide the reader with an idea of where China's actual efforts are focused, and where there is primarily rhetoric and for what reasons. That is the intent of this chapter.

Launch

With the successful launch of AsiaSat-1 in 1990 on a Long March-3 rocket, China joined a short list of countries and organizations involved in the commercial launch business. Manufactured by Hughes for the Asia Satellite Telecommunications, the 1.24-ton satellite is equipped with 24 C-band transponders. Eighty percent of its capacity is utilized for television program transmission, and the rest for communications. The satellite has a life expectancy of 10 years and serves over 2.5 billion people in 30 Asian countries. Asia Satellite Telecommunications was originally a joint venture formed by China International Trust and Investment Corporation (CITIC), British Cable and Wireless PLC, and the Hong Kong Hutchison Whampoa Limited. AsiaSat-1 was insured at launch by the China People's Insurance Company for $120 million. In June 1996, AsiaSat went public. Each of the partners has 23% and 31% public stock, traded in Hong Kong and New York.

The insurance aspect is important to note concerning China's first commercial launch, as in 1997/98 it is among the most important factors concerning China's future in the commercial launch field. China experienced seven launch failures of varying degrees from December 1992 through August 1996,[1] ranging from a launcher exploding before ever reaching the pad, to satellites in low orbits, to catastrophic explosions killing people on Earth. These difficulties resulted in a general loss of confidence in the Chinese Long March launcher family among the commercial satellite vendors and a subsequent increase in insurance rates.

Launch failures are always undesirable, but for the Chinese, theirs came at particularly unfortunate times as within the next 2 years launch contracts will likely be signed for the majority of new global mobile phone networks being developed which rely on low-orbit satellites: e.g., Iridium and Globalstar, backed by Motorola and Space Systems/Loral. The Chinese fear, which is justified, is that they could miss out on the launch contracts for these ventures to a substantial degree because of the lack of confidence and concomitant insurance issues now associated with the Long March boosters. " 'There is a crunch over the next 12 to 18 months,' said Chris Lanzit, satellite business-development director for U.S. based Hughes Corporation, which built ChinaSat. 'There is tremendous growth in the number of satellites being launched and not enough vehicles. By 1998 a whole lot of new vehicles will be available.' "[2]

Insurance is particularly critical because if insurers refuse to insure LM launches China would be forced to withdraw from the lucrative launch market. Without the income derived from the launch market, many of their other space programs financed through launch would have to be seriously curtailed, especially esoteric fields important for their political but not necessarily pragmatic returns, like manned spaceflight. Some analysts have said that " . . . the Chinese programme is rapidly becoming uninsurable . . . "[3] because of its outdated technology, particularly data-handling facilities. Others are not so pessimistic. Rates for insuring a launch on LM did drop from 27% of the amount insured to about 19%, which is still above the cost of less than 17% on Europe's Ariane 4 booster but still considered in the "tolerable" range."[4] After two later accidents though (Intelsat 708 and ChinaSat-7) rates for the LM-3B went up again to between 20–30%. All, however, seem to agree that if the launch difficulties which have plagued the Chinese recently continue and the insurers pull the plug, then China will miss out on not only a valuable opportunity, but likely already planned-for revenue. The Chinese are keenly aware of this dilemma and have made restoring confidence in the Long March launchers their number one space priority, even above development of new launchers. Both a vehicle for delivering small satellite payloads and a vehicle which could be used for planetary exploration or manned spaceflight have been cited as desirable and within Chinese consideration[5] but in both cases development heavily depends on funds derived from the commercial program, so the plans almost inherently become moot without a viable, meaning perceived reliable by the launch market, commercial launcher.

Comparatively speaking, the success rates of Arianespace and Lockheed Martin are usually cited at approximately 95% and the Proton at 93%. China, however, has plummeted to about 80%.[6] With the 18 August 1996 launch failure, China's failure rate was 30%, three times higher than what industry experts consider "reasonable."

Those figures are approximations, for comparative purposes, and can vary because of calculation differences, for example, whether one includes test flights. Further, in the case of China, cautionary warnings have been given in the past about the reliability of the data. Gordon Pike, in a 1991 article where he had assembled a table on Chinese launch failures and successes, stated the following.

> ... these data should be approached with a degree of caution and the reader is advised to research the results of each mission. This is especially true of those missions classified as failures. In the case of the first flights of both the LM-3 and LM-2E the "failure" was of a relatively minor nature (although the consequences were not!) and the essentially successful nature of the launch should be recognized.

The overall point is, be wary of data as meanings behind such may not be clear, deliberately or unintentionally.

> It should also be noted that consistent failure statistics for the military FB-1 and CSS-4 vehicles have not been publicized by the Chinese. Although confirmed by at least one public domain Western source in each case, the figures designated ... reflect the possibility that these sources are incorrect and no such failures occurred.[7]

The importance of that quote lies not only in its statistical warnings, but in the references to the military systems, in that there is considerable design carryover between military and commercial systems. For example, a reorientation problem with LM-2E was traced to a component which had worked successfully in the M11,[8] but did not work in the LM-2E and had to be reworked. Failures of military systems which are not reported in the West may have impact on the commercial program.

China's launch difficulties began in December 1991 when an LM-3 failed to put a DFH-2 into orbit.[9] Then in the spring of 1992, the first attempt to launch Aussat B-1 failed to lift off the pad. There was, however, no damage to the satellite. In December 1992, an LM-2E (predecessor of LM-3B) carrying an Australian telecommunications satellite exploded about 2 minutes into the flight. That was followed by another mishap in April 1994. In that case the rocket exploded in the preparation facility before being taken to the launch pad, with a Chinese weather satellite on board. A November 1994 launch of a Dong Fang Hong-3 communications satellite ended with the satellite lost in space. In that case though, the problem was with the satellite, not the launch vehicle. All of these incidents resulted in serious monetary and technology losses. The truly bad publicity for China extending beyond the space community began with the next failure, in January 1995.

In January 1995, an LM-2E exploded 50 seconds after liftoff. In that case, not only was the Apstar-2 communications satellite it was carrying lost, but falling debris landed on a house, killing 6 and injuring 56 (see Figure 5.1). Then at its February 1996 debut flight, the LM-3B carrying an Intelsat 708 satellite built by Space Systems/Loral blew up in a fireball about 20 seconds after liftoff. The failure was traced to a faulty inertial measuring system, although that information was not provided until some 9 months later, and the system is said to have been subsequently redesigned.

Perhaps what has been most disturbing to foreign users of Chinese rockets, program observers, and the public in general has been the abstruse Chinese attitude and response to the launch failures which resulted in loss of human life. As the LM-3B carrying the Intelsat 708 began to veer off course, the live television pictures were interrupted. Coverage was blacked out completely before it exploded, and as the rocket hit the ground the commentator remarked simply: "This has not succeeded."[10]

Pictures taken of the failed launch of the Intelsat 708 by an Israeli launch observer, smuggled out and released in March 1996, showed damage to a vil-

Figure 5.1 Evidence of damage to one of the buildings at the Xichang launch site from an earlier launch failure. *Courtesy of William Pendley.*

lage near the Xichang launch facility after the accident. He estimated that about 100 villagers died from burns and inhalation of toxic gases. The Chinese first denied casualties and then later admitted to 6 deaths and 23 injuries. Damaged areas included housing outside the gate and a "hotel" just inside the gate where launch support teams are lodged. During my March 1997 visit to Xichang, the "hotel" was observed under repair. The most startling visual image, however, was the large chunk of the very nearby mountain that was blown away as though amputated, a stark reminded of both the vulnerability of the area and the power harnessed for launch.

It was only after the January 1995 and February 1996 accidents that people began to realize the extent of the chasm in safety philosophy between China and other launching nations. "An internal memorandum from the international satellite company known as Intelsat, criticizing a 'blindness towards safety' at China's rocket launch site and describing it as 'pathetically short of world standards,' has added to the aerospace industry's growing concerns about the viability of China's commercial satellite launch capabilities."[11] The memo further says, and is confirmed by those who were at the site, that observers were not permitted to leave the viewing area for 9 hours after the explosion which, it is speculated, gave Chinese authorities time to clean up and remove bodies and fueling assertions that the official casualty reports were underestimated. The problem seems to lie in differing motivations and priorities. Whereas launchers in most nations have both automatic and manual destruct systems specifically intended to keep the launcher from potentially landing on populated areas, the Long March destruct system has been geared toward preventing destruction of the launch pad. Chinese self-destruct mechanisms (manual and auto) were initially fixed with a 20-second delay to give the rocket time to safely clear the launch pad, and at one point, an additional 20-second delay was added. With the LM-3 the delay was reduced to 17 seconds, which takes the vehicle approximately 1.7 km from the launch site before destruction.

Two official reports were done after the Intelsat crash: one by China, only small portions of which have been released, and one by an international panel led by Space Systems/Loral, from which only a synopsis report could be released because it was compiled using U.S. space know-how, the export of which is restricted by the Missile Technology Control Regime (MTCR) internationally and U.S. domestic regulations. Although there have been changes to U.S. technology transfer regulations in the years since the ending of the Cold War, many of the underlying premises behind them have remained in spirit.

The MTCR is actually a policy announced by seven governments in 1987 to limit the spread of nuclear-capable missiles. Originally, MTCR members released guidelines explaining the regime's general principles and an annex defining certain technologies considered desirable to be controlled. Included in those were parameters restricting missile-related exports to those under the capabilities of a 500 kg payload and 300 km range. Concurrently, a wide spectrum of activities was permitted, including provisions for educational exchanges, research programs, and servicing agreements. By 1993, the concerns of the group had broadened to include devices capable of carrying chemical and biological weapons. That change substantially tightened control parameters, since chemical and biological warheads can be placed on small rocket systems that fall well under the previously established 500 kg and 300 km parameters.

Today membership in the MTCR has expanded to 26 governments.[12] Members are those nations which either joined at regime inception or later submitted applications which were subsequently approved. Some other states have claimed to be "adherents" to MTCR principles. Although MTCR states welcome and encourage adherence to regime principles, simple declaration of being an adherent does not necessarily make a country recognized as one, as has been the case with South Africa. The United States has in the past only recognized adherent status after a bilateral accord has been reached with the country in question. Recently Russia and Israel have become recognized as MTCR adherents after the United States reached such an agreement with each country. U.S. recognition as an adherent is particularly critical, as U.S. sanction laws are triggered when non-MTCR participants transfer controlled goods to other nonparticipants. In the case of China, support for adherence was not given before considerable pressure was exerted from the United States, including the threat of sanctions.

Two second-generation Aussat satellites set the precedent for launching foreign spacecraft in China, as that was the first contract signed with the People's Republic of China for the launch of a large Western communications spacecraft. The then Aussat manager for spacecraft procurement took the time and effort to describe the process for the benefit of those who follow.[13] That description is insightful in terms of illustrating attitudes and philosophies which can still come into play if so deemed appropriate by the United States in response to what it considers actions which could contribute to proliferation.

Basically, controls on exports to China emanated from the various licenses required from the U.S. government in relation to the use of Chinese launch services. All licenses were issued by the Office of Defense Trade Controls (ODTC) within the Department of State (DOS). Not only did they cover the export of technical interface data; the export of tooling and electrical connectors for the launch vehicle umbilical wiring; hardware and equipment; but also for items on the Munitions Control (MC) list there were a number of provisos which had to be adhered to. The list below illustrates the tenets which, official or not, many people still feel are supported to some degree or another.[14]

- The Chinese shall only be given data related to the form, fit, and function of the interfaces between the spacecraft and launch vehicles and launch site. Specifically, information may be exchanged related to orbit requirements, launch window, weight, center of gravity, physical envelope, dynamic loading, electrical power usage, interface adaptor requirements, radio frequency plans, safety plans, test flows, separation characteristics, ground handling and test equipment, and flight event sequences.
- What may *not* be transferred to the Chinese is governed by the principle that the spacecraft and its launch program cannot be used as a tool to assist China in the design, development, or enhancement of its satellite, launch vehicle, or missile programs.
- In a more subtle vein, the governing principles behind the licenses also prohibit the transfer of substantive information as to how a program is executed in a typical Western aerospace company. Specifically, test philosophies, organization rationale, industrial relationships, and other procedural issues are not permitted to be transferred.
- Finally, the U.S. government requires that its representatives maintain full visibility of the spacecraft to launch vehicle integration program. In practice this means that U.S. government representatives attend all interface meetings and have visibility of all technical exchanges with China.

During the Clinton administration, Commerce took over from State as the lead U.S. department in the process of issuing export licenses, but safeguards against technology transfer have been and remain fairly comprehensive, targeting not just technology, but know-how. Indeed U.S. military personnel can accompany and guard U.S. satellites prior to launch. Although the technology transfer school of "build high fences around all American technology" for the most part went out with the Reagan administration, replaced by a "high fences in few places, stay ahead in technology" approach, clearly the high fences being imposed with regard to China remain high.

The Chinese have openly expressed bitterness about U.S. refusal to release the external accident report. In the case of the launch failure reviews, the first external accident review the Chinese participated in was after the 1995 Apstar-2 failure. There, two outsiders and one Chinese jointly reviewed the data, but failed to come up with a conclusion that all could agree on, with the split being between the Chinese representative and the two non-Chinese. Together, these two experiences with accident reviews have made launch customers concerned about apparent Chinese hesitancy in being forthright in the review process, or accountability. Actually, from a cultural perspective, the Chinese response has been completely within expected parameters.

Although the Chinese claim to have "fixed" many of the safety problems resulting in the deaths which so horrified the rest of the world, primarily by moving some facilities, the changes are likely not ample. There are villages, fields, and mountain pathways surrounding the Xichang site (see Figures 5.2–5.4), teeming with children and farmers the day I visited. Chances of removing all of them, or of even locating all of them, on any particular launch day are slim at best.

The handling of the accidents has also created a resurgent foreign concern regarding the issue of liability for damage caused by a launch vehicle. The issue generally has been the subject of much debate in the emerging field of space law. While the United Nations Outer Space Treaty of 1967 (Article 7) and the subsequent Liability Convention of 1972 provide *prima facia* definitions of the various responsibilities, they were clearly formulated in an era when the international nature of the modern space industry was not anticipated. There is an Institute for Space Law in Beijing, but these types of matters are apparently primarily the concern of the China Great Wall Corporation because of the business aspects, whereas the Institute for Space Law deals with issues like space debris. Questions on space law issues I raised in China generally drew dismissive or generic answers.

Since the 1995/96 failures, the Chinese have stopped direct video monitoring of their launches, as too potentially embarrassing. Although true, that ac-

Figure 5.2 Road approaching Xichang launch site, showing the rural environment. *Courtesy of William Pendley.*

Figure 5.3 Farms near the Xichang launch site and in the surrounding mountains. *Courtesy of Rob Preissinger.*

Figure 5.4 Farms near the Xichang launch site. *Courtesy of William Pendley.*

tion is not a big confidence booster for outside observers. The American public and much of the world watched Vanguard blow up on the pad at Cape Canaveral in 1957, and *Challenger* in 1986, much to the chagrin of U.S. officials. However, had the United States conducted succeeding launches in secrecy, that would likely not have been viewed by the public as particularly indicative of subsequent confidence about reliability.

What the Chinese hope was the last in their string of misfortunes occurred on 18 August 1996. An LM-3 carrying a Hughes HS-376 satellite called the ChinaSat-7 suffered an incomplete third stage burn which left the satellite in a low orbit. The satellite owner, China Telecom, has been paid $25.9 million by People's Insurance Property Co. as a partial insurance payoff, the rest of the total insured amount of $102 million will come from the other insurers.[15] The result of these accidents on insurance costs has already been substantial. Premiums charged for the failed August launch were stated to be 25% to 30% of the satellite's value. By comparison, companies have paid premiums as low as 15% for launches in the United States.

Concerns about the launch failures are particularly high because three of the last five failures have involved Western satellites, which have left some in the business wondering if there is an integration problem?[16] Publicly, both sides say that integration is not an issue, but both sides also claim a reluctance of their counterparts to share the necessary information to make a complete and valid analysis. Based on the known compartmentalization problems which seem inherent to the Chinese system, integration problems are not a particularly far-fetched conclusion for outside analysts to draw.

Another issue of concern is maintenance. The Chinese are not particularly known for taking a high maintenance approach to buildings and hardware generally, preferring in many cases to just tear things like buildings down and start over. In the space business, however, that approach is simply too expensive. As perhaps the ultimate throwaway society, the United States has to be careful about pointing fingers at others on this issue. But in China the issue is not one of being overly consumptive, it is one of not understanding that certain items are "high maintenance," requiring constant care and upkeep to work properly. In one sense, it is quality control, but it extends beyond the production phase and into usage. Again, this seems to be a cultural factor, evidenced by decay of even relatively new structures in any Chinese city, in factories, and even at "preserved" historical sites. Another consideration in a similar vein is that although Chinese scientists purport to adhere to the ISO 9000 international standards for documentation, some outside analysts question their real commitment to adhering to even their own specifications, saying that the Chinese must constantly, if subtly, be reminded to stick with the planned program.

The Chinese launch manifest has already been impacted by the Long March failures. Denver-based Echostar switched to Ariane to put its Direct-Sat-1 into orbit. That company had used an LM-2E just a month after the AsiaSat-2 lift-off to put its first direct broadcast satellite into orbit (EchoStar-1). Pricing and launch date availability caused one of China Great Wall's best customers, Asia Satellite Communications to shift its Hughes-built AsiaSat-3 to a Russian Proton. The decision will keep AsiaSat on schedule to launch in late 1997; when the decision was made LM was booked until 1998. CITIC, China's largest state-owned overseas investment firm, is a minority partner in AsiaSat. APT Satellite Co. (75% Chinese owned) has remained on the LM schedule, while a Singaporean-Taiwanese satellite venture that was considering China opted for Ariane. Intelsat began outright termination proceedings in April 1996 of its contract to use LM boosters for its launches in 1997/98, and will use Lockheed-Martin Atlas 2A's instead. Argentina's Nahuel-1B satellite, which originally held an optional slot on a Chinese launcher, could be switched to an Ariane 4 or 5. China is a partner in Iridium. Early plans called for China to be first to launch, but now will follow Delta and Proton. The LM manifest for Iridium is for 11 missions (22 satellites) through 2002. CGWIC and Space Systems/Loral signed an agreement in October 1997 to launch the U.S. built ChinaSat-8 satellite on LM-3B. Then in March 1998, Space Systems/Loral contracted with CGWIC for two more launches, tentatively in 2000 and 2001. Cumulatively, this brings CGWIC's backlog to 18 launches.[17]

Recovery from the launch failures has so far proceeded well, but slowly. After ChinaSat-7, there were successful launches of an LM-3A carrying a DFH-3 satellite and an LM-3 with a FY-2, both Chinese payloads. The big test, however, came with the successful August 1997 launch of an LM-3B, the same rocket as slammed into the hillside in 1996.[18] That rocket carried a large, Philippine-owned, Mabuhay communications satellite called Agila 2, built by Space Systems/Loral. Insurance rates for Mabuhay were reported at 22% if successful, 32% if a failure. Obtaining insurance for another LM-3B, successfully launched 17 October 1997, carrying the Apstar 2R, was dependent upon the successful launch of Agila 2. "Consistency" will be the key to convincing the foreign launch market that the Chinese problems have been resolved.

Even prior to August launch, Mabuhay had problems, as the United States challenged it as violating the launch pricing agreement. U.S. trade officials said that China underpriced the Mabuhay Philippine satellite. China disagreed, saying that U.S. trade officials conducting their analysis of the Mabuhay contract used a reference price, the figure deemed representative of the Western standard, that was too high. China conducted a similar analysis and found that the Mabuhay contract did not violate the accord. Theoretically if it is found that the contract was underpriced, reaction could range from nothing, to the United States being less likely to raise the ceiling on the U.S.-China launch agreement, to more punitive measures. Many analysts feel that regardless of whether a pricing violation is found to have occurred or not, it is unlikely that anything will be done about it by the Clinton administration. Politically, the agreement has resulted in bringing China closer to market prices, and the benefits to be lost in the defense and foreign policy areas are not considered worth risking over the Mabuhay deal.[19]

Generally, it must be considered that the more capital Beijing is able to earn through commercial launches, the less imperative earning money through missile sales becomes. Clearly, both will be pursued. Beijing, however, is far more likely to trade missile sales for something else that it wants if hard currency can be earned elsewhere. If capital is not available from commercial launches though, missile sales will likely be pursued far more vigorously.

Comparatively it costs approximately $110 million to carry a 3-ton satellite on Ariane, $100 million on a Lockheed-Martin launcher, and $70 million with China. Russian launch prices are only slightly higher than as China but they are booked up for several years. A considerable amount of the Russian launch market is now occupied with work associated with the International Space Station.

Summary of Launchers[20]

Long March 2C (LM-2C)

With a payload capacity of 2,800 kg, the LM-2C is suitable for many low Earth orbit (LEO) satellite missions and is mainly used for launches of recoverable satellites (Table 5.1). To date, LM-2C has had 14 successful launches in succession including the launch of the Swedish satellite Freja as a piggyback payload on October 6, 1992. LM-2C/SD is a three-stage vehicle developed according

TABLE 5.1 LM-2C Technical Parameters. *Used with permission of China Great Wall Industry Corporation.*

	1st Stage	2nd Stage
Diameter (m)	3.35	3.35
Mass of propellant (t)	143	55
Propellant	N_2O_4/UDMH	
Engine	YF-21 (4 × YF-20)	YF22(Main) YF-23 (Vernier)
Engine thrust (kN)	2962	742 (Main) 47 (Vernier)
Engine specific impulse (N • sec / kg)	2550	2911 (Main) 2834 (Vernier)
Liftoff mass (t)	213	
Overall length (m)	39.925	
Fairing diameter (m)	3.35	

to the Iridium launch mission requirements. It consists of two stages upgraded on the basis of LM-2C and a smart dispenser (SD) developed with flight proven technologies and hardware.

Long March 2D (LM-2D)

The LM-2D is a two-stage launch vehicle, developed on the basis of the first and second stages of the LM-4 (Table 5.2). The propellant for the two stages is N_2O_4/UDMH. The payload capacity for a 22 km low Earth circular orbit is 3700 kg (Table 5.3). To date, three recoverable satellites have been successfully launched by LM-2D.

TABLE 5.2 LM-2D Technical Parameters. *Used with permission of China Great Wall Industry Corporation.*

	1st Stage	2nd Stage
Diameter (m)	3.35	3.35
Mass of propellant (t)	181	37
Propellant	N_2O_4/UDMH	
Engine	YF-21 B	DaYF-20
Engine thrust (kN)	2962	742 (Main) 46.1 (Vernier)
Engine specific impulse (N • sec / kg)	2550	2910 (Main) 2762 (Vernier)
Liftoff mass (t)	233	
Overall length (m)	37.728	
Fairing diameter (m)	2.90 (Type-A)/3.35(Type-B)	

TABLE 5.3 LM-2D Orbital Performance. *Used with permission of China Great Wall Industry Corporation.*

	Parameters	Deviation (3δ)
Perigee altitude	200 km	±5 km
Apogee altitude	200 km	±5 km*
Inclination	60°	±0.2°
Perigee argument	–	±5°

*Semimajor axis

TABLE 5.4 LM-2E Technical Parameters. *Used with permission of China Great Wall Industry Corporation.*

	Boosters (4)	1st Stage	2nd Stage
Diameter (m)	2.25	3.35	3.35
Mass of propellant (t)	4 × 37	187	86
Propellant	N_2O_4/UDMH		
Engine	4 × YF-20	YF-21 (4 × YF-20)	YF-22 (Main) YF-23 (Vernier)
Engine thrust (kN)	4 × 740	2961	742 (Main) 47 (Vernier)
Engine specific impulse (N • sec / kg)	2550	2550	2911 (Main) 2834 (Vernier)
Liftoff mass (t)	460		
Overall length (m)	49.7		
Fairing diameter (m)	4.2		

Long March 2E (LM-2E)

The two-stage LM-2E launch vehicle has first and second core stages similar to those of the LM-2C (Table 5.4). There are four boosters strapped on to the first stage of the launch vehicle, each with a height of 15 meters and a diameter of 2.25 meters. The LM-2E mainly provides low Earth orbit (LEO) satellite launch services and has a LEO capacity of up to 9500 kg. The LM-2E has launched satellites for the Optus, Apstar, AsiaSat and Echostar satellite programs.

Long March 3 (LM-3)

The LM-3 launch vehicle is well-known in the international commercial launch community for the successful launches of the AsiaSat-1 and Apstar-1 and 1A communications satellites. The China Academy of Launch Technology (CALT) developed the liquid hydrogen and liquid oxygen engine for the third stage of this launch vehicle with capability of restarting in a vacuum environment (Table 5.5). The successful development of LM-3 in 1984 makes China the fourth country in the world to launch geosynchronous satellites (Table 5.6). The geosynchronous transfer orbit (GTO) capacity of the LM-3 is 1450 kg.

TABLE 5.5 LM-3 Technical Parameters. *Used with permission of China Great Wall Industry Corporation.*

	1st Stage	2nd Stage	3rd Stage
Diameter (m)	3.35	3.35	2.25
Mass of propellant	144	36	8.7
Propellant	N_2O_4/UDMH		LOX/LH
Engine	YF-21 (4 × YF-20)	YF-22 (Main) YF-23 (Vernier)	YF-73
Engine thrust (kN)	2962	742 (Main) 47 (Vernier)	44.4
Engine specific impulse (N • sec/kg)	2550	2911 (Main) 2834 (Vernier)	4119
Liftoff mass (t)	204		
Overall length (m)	44.56		
Fairing diameter (m)	2.60/3.00		

TABLE 5.6 LM-3 Orbital Performance. *Used with permission of China Great Wall Industry Corporation.*

	Parameters	Deviation (1δ)
Perigee altitude	200 km	±6 km
Apogee altitude	35786 km	±50 km*
Inclination	29°	±0.07°
Perigee argument	179.2°	±0.29°

*Semimajor axis

Long March 3A (LM-3A)

Designed and developed from advanced LM-3 technology, the LM-3A introduces powerful cryogenic third-stage engines, a more capable control system, greater flexibility in the attitude control system, and improved adaptability (Table 5.7). LM-3A has a geosynchronous transfer orbit (GTO) capacity of 2600 kg and can be used for LEO, Sun-synchronous orbit (SSO), and polar orbit satellite missions as well (Table 5.8). The LM-3A's first test flight was successfully conducted on February 8, 1994.

TABLE 5.7 LM-3A Technical Parameters. *Used with permission of China Great Wall Industry Corporation.*

	1st Stage	2nd Stage	3rd Stage
Diameter (m)	3.35	3.35	3.00
Mass of propellant (t)	172	30	18
Propellant	N_2O_4/UDMH		LOX/LH
Engine	YF-21 (4 × YF-20)	YF-22 (Main) YF-23 (Vernier)	YF-75
Engine thrust (kN)	2962	742 (Main) 47 (Vernier)	157
Engine specific impulse (N • sec/kg)	2550	2911 (Main) 2834 (Vernier)	4286
Liftoff mass (t)	241		
Overall length (m)	52.5		
Fairing diameter (m)	3.35		

TABLE 5.8 LM-3A Orbital Performance. *Used with permission of China Great Wall Industry Corporation.*

	Parameters	Deviation (1δ)
Perigee altitude	200 km	±10 km
Apogee altitude	35786 km	±40 km*
Inclination	28.5°	±0.7°
Perigee argument	179.8°	±0.2°

*Semimajor axis

Long March 3B (LM-3B)

The LM-3B launch vehicle was designed with an LM-3A launch vehicle as the core stage with four liquid boosters strapped on to the first stage (Table 5.9). The core stage of LM-3B is identical to the LM-3A's except that the stage tanks have been extended and reinforced, the fairing has been enlarged, and the control and telemetry systems include minor modifications to accommodate the strap-on boosters. LM-3B is capable of launching a payload of up to 5000 kg into geosynchronous transfer orbit (GTO) as well as performing missions to other orbits. In addition, LM-3B is also capable of accomplishing payload attitude adjustments, reorientation and spin-up requirements, and dual or multiple launch requirements.

TABLE 5.9 LM-3B Technical Parameters. *Used with permission of China Great Wall Industry Corporation.*

	1st Stage	2nd Stage	3rd Stage
Diameter (m)	3.35	3.35	3.00
Mass of propellant (t)	171.8	49.6	18.2
Propellant	N_2O_4/UDMH		LOX/LH
Engine	YF-21 (4 × YF-20)	YF-22 (Main) YF-23 (Vernier)	YF-75
Engine thrust (kN)	2962	742 (Main) 47 (Vernier)	157
Engine specific impulse (N • sec/kg)	2550	2911 (Main) 2834 (Vernier)	4286
Liftoff mass (t)	425.5 (LM-3B)		
Booster	4 (LM-3B)		
Overall length (m)	54.838		
Fairing diameter (m)	4.00/4.20		

Long March 4 (LM-4)

LM-4 is a three-stage launch vehicle (Table 5.10). The first and second stages of LM-4 were developed on the basis of LM-3. The third stage is newly developed. The propellant for all three stages is N_2O_4/UDMH. The payload capacity for 900 km SSO is 1650 kg (corresponding to a single burn of the third-stage engine) and 2800 kg (corresponding to two burns) (Table 5.11). To date two meteorological satellites have been successfully launched by LM-4.

Satellites

Communications

Linking the rural and inland regions of China to the more coastal cities has already been cited as a constant consideration, indeed priority, of the Chinese modernization program. Remembering the many historical examples of uprisings in those regions as a result of feeling cut off from the concerns of central government, China's leaders today want to avoid that mistake of the past. One of the most curious sights traveling in China is the surprisingly high number of people who walk around talking on their cellular phones, which are even made available for rental through the tourist hotels. The reason becomes clear after only a short time; outside the hotels, phones for public use are few and far between.

TABLE 5.10 LM-4 Technical Parameters. *Used with permission of China Great Wall Industry Corporation.*

	1st Stage	2nd Stage	3rd Stage
Diameter (m)	3.35	3.35	3.00
Mass of propellant (t)	182	35.4	14.3
Propellant	N_2O_4/UDMH		LOX/LH
Engine	YF-21B	DaFY-20	YF-40
Engine thrust (kN)	2962	742 (Main) 46.1 (Vernier)	100.85
Engine specific impulse (N • sec/kg)	2560	2910 (Main) 2762 (Vernier)	2971
Liftoff mass (t)	249		
Overall length (m)	45.776		
Fairing diameter (m)	2.90 (Type-A)/3.35 (Type-B)		

TABLE 5.11 LM-4 Orbital Performance. *Used with permission of China Great Wall Industry Corporation.*

	Parameters	Deviation (3δ)
Perigee altitude	900 km	–40 km
Apogee altitude	900 km	±40 km*
Inclination	99°	±0.1°
Perigee argument	0°	±0.003°

*Semimajor axis

Still, currently only about 3.76% of China's population has phone service.[21] China has been using space links as its chief means of communication between regions since 1984. By mid-1986, five Chinese cities, Beijing, Lhasa, Urumqi, Hohhot, and Guangzhou, had domestic public communication lines. As of 1995, construction was completed for 10 domestic and international optical trunks, between: Beijing-Hankkou-Guangzhou, Beijing-Huhehot-Yinchuan-Lanzhou, Hangzhou-Fuzhou-Guizhou-Chengdu, Beijing-Shenyang-Harbin, Beijing-Taiyuan-Xian, Hankou-Chingqing, Jinan-Qingdao, Nanjing-Hangzhou, Urumqi-Yining, and Sino-Korea lines.[22]

China's post and telecommunications (P&T) industry is growing at a high rate. The Chinese public communications network virtually completed the transition from manual to automatic operation and from analog to digital technology in 1995. The growth rate of the total turnover of P&T services exceeded that of the Chinese GNP for the 11th consecutive year.

The total turnover of P&T services in 1996 was $11.78 billion and the revenue was $11.77 billion, depicting a growth rate of 42.2% and 43% respectively over the prior year. An investment of $11.78 billion was made in P&T fixed assets, an increase of 27% over the prior year, increasing the total P&T fixed assets to $31.2 billion.[23] Urban and rural telephone exchange capacity quadrupled in 5 years, reaching 85.1 million lines. Also, 180,000 mobile channels were opened, with mobile networks covering most cities and counties and some townships.[24] The growth of the telephone market in China, particularly regarding mobile telephones, has surpassed all predictions.

As stated prior, this puts foreign businesses in a curious dilemma. In order to get into the burgeoning Chinese market quickly, they are forced to accommodate the traditional Chinese manner of operation, through *Guan Xi*, which perpetuates a business mode of operation which is not particularly in their favor. What that will mean for the long term remains unclear.

Although CASC reports that it has developed and manufactured more than 30 various kinds of satellites for science and technology experiments, most fall into three categories: low orbit recoverable remote sensing satellites; geosynchronous communication-broadcasting satellites; and Sun-synchronous (or polar orbit) meteorological satellites. The Dong Fang Hong communication satellite already referenced is the workhorse of the stable (Table 5.12).

Also as noted prior, neither of the only two DFH-2 satellites ever built are currently operational. DFH-2 had a relatively low capacity, with only four transponders, and provided only a short-lived, partial solution to Chinese communication needs. The first DFH-3 satellite had a fuel leak and was never put into service. The second DFH-3 was launched in May 1997 and is rumored to have some technical problems which render it less than fully functional. Both DFH-1 and DFH-2 were completely indigenous in design and manufacture. DFH-3, however, relies heavily on foreign-sourced parts and was basically a Western design assembled in China. This was a new approach by which the Chinese hope to eventually develop larger, more powerful satellites. The result, however, has been less than satisfactory.

Some analysts have compared difficulties with the DFH-2 and DFH-3 satellites with the still ongoing struggle between the Civil Aviation Association of China (CAAC) to buy Western-built commercial aircraft, primarily from Boeing, and the Aviation Industry of China (AVIC, formerly the Ministry of Aviation) desire to build aircraft in China. In the satellite-world parallel, the MPT wants Western satellites, while CASC wants to be the MPT's supplier. MPT has a long history of disappointments with CASC: the DFH-2 which

TABLE 5.12 Dong Fang Hong-3 Technical Parameters. *Used with permission of China Great Wall Industry Corporation.*

Parameter	Value
Dimension (mm)	2000 × 1720 × 2200
Liftoff mass (kg)	2260
Dry mass (kg)	945
Stabilization	3-axis
Payload	24 C-band transponders 6 medium power, 16 W 8 low power, 8 W
Frequency (GHz)	uplink 5.926 ~ 6.425 downlink 3.700 ~ 4.200
Onboard antenna	2 m, shaped beam orthogonal linear polarization
Power output of solar array	BOL (at equinox) 2049 W EOL (at equinox) 1688 W
Operating lifetime	8 ~ 10 years

failed to reach orbit in December 1991 due to an LM-3 third-stage problem; a DFH-3 failure in late 1994 attributed to leaked fuel; and the ChinaSat-7 failure to reach orbit because of an LM-3 third-stage problem, again. This policy issue will be an important one in the future.

Satellites are utilized for purposes beyond private communication also. Using satellite television as a tool for education has been a part of China's modernization drive for many years. The problems faced were and remain arduous, involving not just the students but the training of teachers as well. In 1988 it was stated that, " . . . there are 8 million teachers in elementary and middle schools and 2.4 million of them require training to be able to teach all levels in China's nine-year compulsory education. There are also 2.6 million people needing further on-the-job education, and training in a specialization that will serve the country's current technological and economic development."[25] Teachers are a valuable commodity in China. At that time, China had already set up a program in 1986, called China Education Television (CETV), as part of the second phase of the "Leasing for Transition" program. The program was broad based, including teacher training, television universities, and adult basic education aimed at working adults who wished to learn in their free time. Already it is estimated that more than 2 million people received university and technical education through TV transmitted courses, as well as management and technical training.[26]

Remote Sensing

The Chinese launched two land survey satellites in 1984 and 1986, which took more than 3000 land images, each one covering 32,000 square kilometers of the Earth's surface.[27] I was proudly told in China that Chinese remote sensing activities have expanded rapidly over the past 15 years and currently involve more than 460 institutes and agencies, and upward of 10,000 researchers. Although on one hand this could be indicative of a flurry of activity, it is also surely indicative of the need to keep people employed, with the likely corollary of inefficiency which is seen in other sectors. It is thought that the Chinese Academy of Sciences and the Shanghai Institute of Technical Physics are heavily involved with remote sensing, the Shanghai Institute of Technical Physics building optical sensors.

The main satellites used by China for land management, vegetation monitoring, cartography and other such applications are Landsat (United States), Spot (France), JERS-1 (Japan), and its own FSW series of recoverable capsules. The FSW series satellite are recoverable satellites which use the LM-2C and LM-3D as launch vehicles. China has successfully launched and recovered 14 satellites since 1975.[28] The camera system carried by China's FSW-2 type spacecraft can be used for military reconnaissance and remote sensing. The camera carries 2000 meters of film and has a resolution capability of at least 10 meters.[29] The FSW-2 was launched in October 1996 using a LM-2D from the Jiuquan launch site.

China is currently developing its first Earth resources remote-sensing satellite, the China-Brazil Earth Resources Satellite (CBERS), in partnership with Brazil. CBERS will feature a CCD camera (with 20 m resolution), a multispectral infrared scanner (160/180 m), and a wide-field imager (258 m).[30] The Shanghai Institute of Terrestrial Physics has been a key player working with Brazil on CBERS.

China is also expected to build a new radar remote sensing satellite for launch in about 2002 for both civil and military uses. Apparently, it would be comparable to Canada's Radarsat and the European Space Agency (ESA) Earth Resource Satellite vehicles (ERS-1, ERS-2). The comparisons are not coincidental. The technology sought to be used is beyond what China has now, and outside participation would be necessary. Both GEC-Marconi and DASA are said to be interested in the potentially lucrative deal, with estimates of the satellite in the $250 million class.

Meteorology is also a major area of interest to China, with its expansive land covering and need to warn its population about meteorological events and aftermaths such as typhoons and flooding. Two FY-1 satellites preceded the first FY-2, launched in 1997. An earlier FY-2 was the satellite which exploded in the integration hall at Xichang in April 1994. The Chinese have announced that they are also planning development of a more advanced geosynchronous-orbit weather satellite system, as well as enhancement of their polar-orbit spacecraft.

Manned Spaceflight

Manned[31] spaceflight is clearly the ultimate prize in the prestige race for which space has long been a tool.[32] The Chinese have in the past released pictures of astronauts in training and made public their ambitions to put men into orbit.[33] They continue today to allude to a manned program, though more enigmatically. At the Zhuhai airshow in November 1996, for example, the Chinese displayed a mock-up of a space shuttle. Whether there is anything more to the shuttle program than a papier-mâché mock-up is questionable. Indeed the status of the entire manned program is currently one of the more ambiguous areas of Chinese intent.

Reports on Chinese manned space activities range from those citing "Project 921" which has allegedly been in existence under the Defense Ministry since 1992, to those which say Chinese manned spaceflight ambitions will be based on the LM-2E launcher, capable of placing 9.2 tons payload in LEO,[34] and its experience with FSW recoverable capsules. The latter scenario is sometimes accompanied by reports of a space plane to be launched by a modified LM-2E and manned by up to five astronauts. If accurate, the world will know relatively soon as two tests or demonstration flights are supposedly planned for 2000–2001 before flying with two cosmonauts in 2001–2002.[35] These plans have recently been reiterated with the announced development of an upgraded LM-2E for heavy lift, including manned spaceflight, referenced as the LM-2E(A). It would be launched from Jiquan. Also, at an April 1998 meeting with French officals, Chinese officials talked about a manned spacecraft resembling the Russian Soyuz and based on FSW technology, to be lofted by the LM-3B.[36]

In other forums or to reporters, China has also seemingly discussed establishment of its own space station by 2020. Sometimes specifics are given:

> China's manned spacecraft will be roughly comparable to the two-man U.S. Gemini craft first launched in 1964, although its size could be closer to Russian Soyuz . . . Like Gemini and Soyuz, the rudimentary Chinese spacecraft will be a wingless ballistic reentry vehicle. But it will not have the rendezvous, docking or flight endurance of either Gemini or Soyuz.[37]

The problem seems to be, however, rarely having the same set of specifics, even suggested specifics, officially given twice.

In Beijing, I asked one Chinese gentleman with long experience in the space program his opinion about Chinese intentions concerning a manned space program. He laughed. He said that in the 1970's there had indeed been a serious manned spaceflight program, but that the money had run out and that all those who had been astronauts-in-training were currently working in the basement of the same building he is, old men ready to retire. Another person I asked, comfortable in English and well acquainted with Western ways, immediately switched to Chinese. She told the interpreter that I should not ask other people about that topic until they felt comfortable with me as it would scare them off. Official Chinese statements of intent for the future of their space program always include a phrase similar to "including manned spaceflight," and their eventual intent in that direction is clear. At the 1996 IAF meeting in Beijing, CASC Vice-President Wang Liheng stated that China's plans for the future include a manned spaceflight technology research program as well as other projects such as lunar and planetary probes; a data-relay satellite; new telecom and remote observation probes; and new launchers. But all of these are very ambitious and require substantial funding, hence their typically oblique reference only.

When time specifics are given, apparently they are fungible and not well publicized even within CASC. A Chinese publication stated that "China expects to be an important player in space by the year 2010, including manned space flight."[38] *Aviation Week & Space Technology* reported that "China is planning to launch two astronauts into space in 1999 to celebrate the 50th anniversary of the founding of the Chinese communist state."[39] When I brought these statements to the attention of one of the CASC vice-presidents in Beijing, he dismissed them saying he was not aware that any date had ever been given to or reported in the U.S. press.

It has also been suggested that China is currently engaged in manned spaceflight technology in partnership with Russia.[40] Although specifics on those arrangements vary considerably, they are based on intergovernmental agreements signed between the Russian and Chinese space agencies in 1992 and 1994. It is clear that the Chinese have long placed their students at the prestigious Moscow Aviation Institute (MAI)[41] and two individuals, military pilots Wu Tse and Li Tsinlung, took a general training course at Russia's cosmonaut training center in 1996–97. The Chinese have also indicated that a Sino-Russian cosmonaut exchange is possible once they have a piloted spacecraft in orbit. Russian sources have said that it will be the Russians who build a controllable spaceship in China and provide active cooperation toward China launching a 20-ton spacecraft into orbit. Anything is possible. What becomes relevant here, however, is what is financially probable.

The Chinese attended the training center on a contract basis with the Russians, who had contingents from Germany, France, and the United States there at the same time. Clearly, the Russians are interested in currency generation, a fact not lost on proliferation analysts who fear Sino-Russian arrangements could extend to technology sales as well. These fears will likely be justifiably heightened if commercial space revenues for both parties are not high enough to offset "restrained" arms sales.

The money for an indigenous Chinese manned space program must come from commercial launches. From within these precious commercial profits, the prestige to be garnered from manned spaceflight is in serious competition with more pragmatic needs, like communications and remote sensing capabilities. Some though have suggested other than indigenous routes, beyond the Russians. Indeed in 1996 it was suggested that the U.S. Space Shuttle be used to rescue the ChinaSat satellite stranded in a useless orbit in August 1996 when the LM-3 third-stage engine quit firing about 48 seconds early after a planned restart failed.[42] The suggestion was particularly peculiar from an economic perspective: using the Shuttle, with its launch cost, conservatively, of $400 million to rescue a commercial satellite with a combined booster/spacecraft insured value outside China of $102 million. Obviously, there was a diplomatic angle being considered. Although it was recognized that the flight could represent a new diplomatic opening to China, possibly including the flight of a Chinese astronaut on the shuttle, the White House was apparently cool on the idea.[43]

Another set of scenarios seems to revolve around the fact that China is the only major space power not engaged in the International Space Station (ISS). This scenario says that the activity level, real or reported, is kept just high enough to posture China for future consideration for inclusion on the ISS. Officially, Chinese officials say that China is interested in ISS participation, but that the time is not yet right because they have neither the funds nor the technology to contribute.

Generally, once beyond the rhetorical stage, most of those in China who are actually involved in space-related activities are against manned spaceflight activities in the near term, saying it takes resources away from more pragmatic programs like communications and remote sensing. But prestige has certainly proven to be a serious impetus for costly space activities in other countries, and such could prove to be the case in China too. The timeline seems to be the real question.

Space Science

Although in all the public speeches to international audiences space science is included in the list of current activities and plans for the future, the reality is that it is totally government funded and hence, with pragmatism being the ruling order of the day, space science has taken a back seat to other fields with more immediate applications potential, like meteorology and telecommunications. This is not surprising as it is the same situation as is being faced in other space programs, taken to a harsher level. Work that is being done is university centered. State Science and Technology Councilor Song Jian is trying very hard to increase the proportion of government spending in research and development to 1.5% of the gross domestic product from its present 0.6%. Of that, one-tenth would go to basic research.[44]

Although there is certainly recognition in China that in the long term, China's future and ability to bridge the technology gap will depend on its proficiency in basic research, right now the balance is swinging in favor of short-term goals. The Chinese certainly cannot be faulted for this, as the United States and most other countries are suffering from the same malaise, though not to the same degree. Indeed it has just been within the last 10 years that the Japanese have been able to afford to move more of their efforts from applications to basic research.[45]

Progress now is primarily in those areas which are not particularly capital intensive. There is an observatory northeast of Beijing, for example, that does world class work monitoring the magnetic fields on the Sun's surface. Even there, however, emphasis is not only on the science returns, but on the potential for selling copies of the telescope filter which one of the scientists at the facility developed. CAST Vice-President Ma Xingrui has stated, however, that "a small exploration vehicle to the Moon" is being studied.[46]

Part of the problem for the long term also lies in the impact of the science brain drain that China has, and continues to experience. Of the approximately 270,000 Chinese who have gone abroad to study, it is estimated that only about 90,000 (those from rich, privileged families) will return. Those who do stay, or return, are often frustrated on a variety of levels. Not only is there little money to work, salaries are very low (in academics, approximately $40–50 a month), and opportunities few. Indeed when opportunities do come up for paid appointments abroad or even travel abroad, they are most often given to mid level or senior officials. These are given as rewards for patronage, persistence, and a variety of other reasons. A Chinese space scientist I shared an office with for several months in 1993 while we were both working in Japan told me he could earn enough in 1 year to entirely support himself and his family in China after his retirement.

International Cooperation

In every speech that is made regarding plans for future Chinese space activities, among those items at the top of their priorities list is the desire to increase Chinese involvement in international ventures. Like other countries, after the rhetoric in support of mutual goals and goodwill is finished, their reasons for wanting to increase their participation are pragmatic. Partners bring technology and money to the table, both items the Chinese are seeking. Indeed one of the biggest problems they have had is that relative to many countries or businesses that China wants to cooperate with, they have little to offer of either. What they do have, however, is the largest emerging market in the world. The clout that seems to come with that should not be forgotten or underrated.

Other spins on the newfound Chinese eagerness to cooperate with other countries on space ventures include that it is "political spin control" to cloud the damage suffered to the Chinese program after the 1996 launch failures.[47] Others say that the loud, repeated and broad pronouncements that were made at the IAF meeting in Beijing in October 1996 on the subject were posturing for Science Advisor Song Jian's scheduled talks in Washington with U.S. Science Advisor John Gibbons later that month.

China has been anxious to expand its cooperation opportunities with the United States beyond the small number of space science projects already underway. The United States has been reluctant because of technology transfer concerns and regulations, as well as political pressure from those who want to take a tougher line with China. U.S. businesses, however, have also been pressuring the administration to allow them to respond to the tremendous Chinese market potential. "The market is huge but U.S. companies have a tremendous disadvantage compared to the Europeans, because of their ability to offer the kind of technology transfer packages that are restricted under U.S. law . . . We have to find a better way to 'cooperate from a Chinese perspective.' "[48] This according to Thomas J. Dwyer, vice president for business development with Lockheed Martin. Unfortunately, the ways that the Chinese are looking for may be short-term advantageous for business but have potentially clouded long-term consequences.

In January 1997, France and China began exploring a number of cooperative efforts in space, primarily focusing on commercial Earth observation and joint space science missions. On 16 May 1997 the Chinese and French governments agreed to a broad cooperation accord on space research and satellite construction. The agreement came as part of a bilateral meeting in Beijing, believed to include Earth observation and space science missions. Earlier talks about launch vehicle cooperation were not mentioned.[49]

China also signed a broad space agreement with the Ukraine in June 1997. Specifics have not become known. Agreements with countries such as the Ukraine and Russia are particularly worrisome because their financial situation and the need to prop up their own missile industries make them amenable to selling technology which is otherwise not available to the Chinese. For example, there is concern about attempts to purchase SS-18's from Russia.

"Russian technology transfers could facilitate China's development of advanced cruise missile weapons, and one has reason to question whether China can be persuaded to forego exporting them, the MTCR notwithstanding."[50] China says it is interested in purchasing SS18's to improve their space launch systems, though for what purpose is unclear. It is clear, though, that it would likely not be for exploration or manned spaceflight in the near future, not with other more pragmatic issues to tackle first.

As mentioned earlier, China is currently developing a remote-sensing satellite, CBERS, in partnership with Brazil. Interesting that on this project some of the most troublesome problems that came up were over Chinese concerns over unwanted technology transfer to Brazil and the legal arrangements.[51] Although the CBERS project inception can be traced back to 1985, with a launch date initially scheduled for December 1992, the project almost immediately slipped behind schedule due to difficulty with funding on the Brazilian side stemming from political turmoil in Brazil generally, and lack of support for the project specifically. Indeed it was not until 1992 that a secure funding commitment in Brazil could be obtained. Since then, estimates for launch on an LM-4B have ranged from early 1998 to the year 2000, with the earlier estimates likely now being the more accurate.

As originally written, the project agreements state that China would be responsible for 70% of the total project cost, and 30% was to be Brazil's responsibility. These figures were readjusted after the total project cost was more accurately calculated and ambiguous legal responsibilities interpreted to the extent that for the first of the CBERS satellites, the project split is about 60/40 China/Brazil, with China providing most of the space equipment and optics, and Brazil providing the cameras and ground equipment. For the second satellite, the split is expected to be more 50/50. "However, both parties have not, so far, redefined this matter through another agreement,"[52] at least in part because of the legal interpretation problems encountered with the prior agreements.

The difficulties encountered with the legal agreements are important for several reasons. First, "by the time (the two Sino-Brazilian satellites) are in orbit, they will provide the most significant example of South-South cooperation in the whole planet."[53] Second, the difficulties encountered are illustrative and likely indicative of ones that might be anticipated in the future dealing with countries lacking a mature legal system.

> Agreements must be honored . . . Nevertheless, they are not always duly respected. This failure is critical to multilateral as well as to bilateral agreements. It can signify frustration, breaking off an entire chain of understandings from which concrete results could be expected, and endangering the benefits, not only for the countries

> concerned, but also for international relationships as a whole. It is especially crucial with reference to agreements in high technology fields.
>
> Of course, bilateral agreements, as well as laws of reciprocity, are relatively strong, being laws of limited partnership, whose members' common ends are a set of mutual interests. It is usually easier to comply with bilateral agreements. Yet, it does not necessarily happen this way, even when the agreement deals with extremely important affairs.
>
> In fact, agreements, including bilateral ones, may have different levels of respect. Sometimes they are implemented, but not exactly as they were established, and only some degree of compromise has been reached. Quite frequently, parties comply with some of the obligations, but not others, due to some new difficulty or reasoning, leaving it clear, however, by their actual deeds, that they are not giving up the entire agreement.[54]

Legal interpretations can be expected to be made in whatever way most benefits the country making the interpretation.

Because interpreting for one's own benefit has always been true, a certain degree of specificity is desirable in contract law. However, referencing the launch contract between Brazil and China, "Articles 10 and 13, on delicate questions such as 'Defaults' and 'Disputes' were written in poor legal terms, certainly because it was intended that the Agreement be as cordial as possible."[55] Clearly, cordiality and diplomacy must be balanced with sound legal practices, or the resulting legal documents will have little meaning.

The Chinese are particularly anxious to enter into joint ventures with foreign companies, for the multiple returns brought. Outside of Chengdu, I went through a factory where the Chinese were making nose cones for McDonnell-Douglas planes. Obviously, the benefits to China involve both technology and jobs.

Another example involves the EuraSpace GmbH joint venture between Daimler-Benz Aerospace of Germany and CASC. EuraSpace will produce the Sinosat series of communications satellites for Chinese domestic use, with DASA providing the antennas and attitude control systems for the DFH-3 satellites. Created in 1994, the initial investment in the company was reported as $4.4 million, with the first Sinosat satellite to be launched in early 1998. That satellite will replace the DFH-3 that failed early in orbit due to an onboard defect. DASA officials estimate that a dozen satellites could be built through this joint venture arrangement, which also includes Aerospatiale of Paris. In return for the technology transfers from the European firms, they will receive satellite construction contracts from China. There are also discussions that EuraSpace will build remote sensing satellites.

The technology tranfer involved in these ventures has been a source of speculation if not concern for some analysts. "Momentum wheels commercially available from European sources have performance characteristics useful for imagery satellites. The German firm Teldix is exporting its DR50 momentum wheel to China for use on the DFH-3 COMSAT."[56] Thus, the technology for these communications satellites could be diverted and utilized in reconnaissance applications as well. Clearly, however, the characteristics of a remote sensing system capable of high resolution imaging are technically demanding. They include large optical trains, precise attitude control and stabilization of the spacecraft, and high data-rate transmission from the spacecraft to the ground. These technical characteristics easily distinguish spacecraft and sensors from ones that are less capable.

A second joint venture is called Com Dev Xi'an. It is between Com Dev International, a Cambridge, Ontario, firm specializing in cellular radio components, satellite components, and satellite ground stations, and Xi'an Institute of Space Radio Technology (which is under CASC). Created in March 1996, Com Dev Xi'an will build onboard satellite electronics and ground facilities. This joint venture is particularly aimed at developing satellite electronics for sale outside China. China has also been negotiating the acquisition of two Radarsats from Spar Aerospace for $600 million.[57]

Conclusions

China has made remarkable accomplishments in the space-related endeavors, with little assistance to date from the outside. It has demonstrated a range of capabilities over a broad spectrum of activities. Obviously, the Chinese have placed more emphasis and had more success in some areas than others, but one might well conclude that they are bounded by resources and internal obstacles rather than technical ability. Considering all the technological starts and stops, and political hurdles that they have endured and overcome, their achievements take on an even more spectacular hue.

Endnotes

1. Eight failures if the November 1995 launch of AsiaSat-2 is included. In September 1996 Asia Satellite Telecommunications Co. (AsiaSat) sought $58 million from launch insurance providers citing damage to the satellite believed to have been incurred by a rough ride to orbit on a Long March vehicle. Patrick Seitz, "AsiaSat Seeks $58 Million on Insurance Claim," *Space News,* 16–22 September 1996, 3. The allegations have been disputed, however, and whether proven or not remains to be seen. See: Bansang W. Lee, "No Presumptions," Letter to the Editor, *Space News,* 9–15 June 1997, 12, 13.
2. Simon Fluendy, "Up in Smoke," *Far Eastern Economic Review,* 5 September 1997, 69.
3. Mark Ward, "Exploding China's dreams," *Interavia,* 16 March 1997, 15.
4. James R. Asker, *Aviation Week & Space Technology,* 15 January 1996, 43.
5. Zhang Xinzhai, "The Achievements and the Future of the Development of China's Space Technology," *Aerospace China,* Summer 1996, 25; Stefan Barensky, "China Un-

veils Heavy Launcher, Manned Spacecraft," *International Space Industry Report*, 23 April 1998, 1.

6. Zhang, 25.
7. Gordon Pike, "Chinese launch services: a user's guide," *Space Policy,* May 1991, 108–109.
8. The M11 is a solid propellant fueled, second generation ballistic missile with a 300 km range and 500 kg payload capacity. R&D was begun on it in 1985 and a photograph of it was displayed at an exhibition in 1988. Lewis and Hua, 1992, 11–12.
9. See: *Jane's Space Directory, 1996–97*, ed. Andrew Wilson, for a complete listing of launches and outcomes.
10. Ward, 15.
11. Elisabeth Tacey, "Chinese rocket site 'blind to safety,' " *Nature,* 13 June 1996, 542. The Chinese attitude is certainly not new; foreign awareness and recognition of it, however, are. Indeed in the 1970's when the Chinese were examining the issue of missile survivability, they considered a proposal to store missiles in civilian houses with removable roofs. Lewis and Hua, 1992, 24, ft. 26.
12. It ought be noted that membership increased substantially with the ending of the Cold War and the advent of the Gulf War; 13 members joined between August 1990 and June 1993.
13. Gordon Pike, "Chinese launch services: a user's guide," *Space Policy,* May 1991, 103–115.
14. Pike, 113.
15. "ChinaSat Payoff Begun," *Aviation Week and Space Technology,* 14 October 1996, 33.
16. Ward, 15.
17. Stefan Barensky, "Loral Orders Two More Chinese Launches," *International Space Industry Report*, 26 March 1998, 3.
18. Agila 2 had to expend extra fuel after launch to move 3000 kilometers to the correct orbit. "Agila 2 Uses Extra Fuel to Reach Proper Orbit," *Space News,* 8–14 September 1997, 2.
19. Warren Ferster, "Great Wall-Loral Contract Raises Price Questions," *Space News*, 12–18 August 1996, 1, 19.
20. From: Overview of the Chinese Civil Space Presentation, given by Baosheng Chen, Washington, DC representative of the China Great Wall Industry Corporation, International Space University 1997 Summer Session, Houston, TX, 22 July 1997.
21. *The APT Yearbook,* 1997, 307.
22. *The APT Yearbook,* 1997, 308.
23. *The APT Yearbook,* 1997, 308.
24. *The APT Yearbook,* 1997, 309.
25. Wu Guoxiang, "China's space communication goals," *Space Policy,* February 1988, 43.
26. Presentation by Baosheng Chen, "Overview of the Chinese Civil Space Program," Washington, DC, 3 June 1997.
27. Zhu and Xu, 1997, 70.
28. Further, in August 1987 and August 1988, two FSW satellites were used as a microgravity test platform for Matra of France and the German Aerospace Research Establishment, respectively.
29. Craig Covault, "Chinese Manned Flight Set for 1999 Liftoff," *Aviation Week & Space Technology,* 21 October 1996, 22.
30. Technical specifications for CBERS can be found in *Jane's Space Directory 1996–97*, 382.

31. The term is used generically for "human" spaceflight. References by the Chinese are not as careful about being politically correct as in the West though, so I will use the gender-specific term formerly used in the West and still used in China.
32. See: Roger Handberg and Joan Johnson-Freese, *The Prestige Trap,* (Dubuque, Iowa: Kendall-Hunt Publishing, 1994).
33. Ward, 14; Mark Williamson, "Chinese Space Show," *SPACE & Communications,* Nov/Dec 1996, 29.
34. Zhang Xinzhai, "The Achievements and the Future of the Development of China's Space Technology," *Aerospace China,* Summer 1996, 22.
35. Theo Pirard, "Space Business Around the World," *Spaceflight,* September 1997, 294.
36. Stefan Barensky, "China Unveils Heavy Launcher, Manned Spacecraft," *International Space Industry Report,* 23 April 1998, 1,6.
37. Craig Covault, "Chinese Manned Flight Set for 1999 Liftoff," *Aviation Week & Space Technology,* 21 October 1996, 22.
38. Wu Bian, "Space Industry Promotes Modernization," *Beijing Review,* 6–12 January 1997, 14.
39. Craig Covault, "Chinese Manned Flight Set for 1999 Liftoff," *Aviation Week & Space Technology,* 21 October 1996, 22.
40. Pierre Langereux and Christian Lardier, "Launch setbacks fail to dent China's space ambitions," *Interavia,* December 1996, 47–48.
41. As is the Chinese practice elsewhere, sometimes the "students" are not young and training for their careers, but mid and senior level officials. The placement of the older "students" seems to allow the gathering of technological information for utilization in the near term, as well as providing them with the career "perk" of travel abroad, which sometimes include a stipend or allowance above what they would receive inside China. Younger students are sent also, in conjunction with long-term posturing for the future. The problem is when they don't come back.
42. The LM-3 profile for the third stage is two burn, or thrust, phases, separated by a coast phase. Restart was a normal procedure; however, in this case fuel leaked out causing early termination.
43. Craig Covault, *Aviation Week & Space Technology,* 23 September 1996, 21.
44. Fisher, 39.
45. Joan Johnson-Freese, *Over the Pacific: Japanese Space Policy Into the 21st Century,* (Dubuque, Iowa: Kendall-Hunt Publishers, 1993).
46. *International Space Industry Report,* 26 March 1998, 8.
47. Craig Covault, "China Seeks Cooperation, Airs New Space Strategy," *Aviation Week & Space Technology,* 14 October 1996, 29.
48. Covault,, 14 October 1996, 31.
49. *Space News,* 19–25 May 1997, 2.
50. Dennis M. Gormely and K. Scott McMahon, "Proliferation of Land-Attack Cruise Missiles: Prospects and Policy Implications," *Fighting Proliferation, New Concerns for the Nineties,* ed. Henry Sokolski (Maxwell AFB, AL: Air University Press, 1996) 156.
51. See: José Monserrat Filho, "Brazilian-Chinese space cooperation: an analysis," *Space Policy,* May 1997, 153–170.
52. José Montserrat Filho, 1997, 160.
53. José Montserrat Filho, 1997, 155, citing Celso Luiz Nunes Amorim, "Por que a China?" *Revista Brasileira de Technologia,* (review published by National Council for Scientific

and Technological Development—CNPq of the Ministry of Science and Technology), Vol. 19, No. 8, August 1988, 48–49.
54. José Montserrat Filho, 1997, 155–156.
55. José Montserrat Filho, 1997, 160.
56. Steve Berner, "Proliferation of Satellite Imaging Capabilities: Developments and Implications," in Sokolski, 110.
57. Pierre Langereux and Christian Lardier, "Launch setbacks fail to dent China's space ambitions," *Interavia,* December 1996, 46.

Chapter 6

The Future: Analysis, Recommendations, and Conclusions

Introduction

In 1991 at a meeting of representatives from several of the world's space agencies in Prague, Czechoslovakia, which I was attending as an observer, individuals took turns briefing their international colleagues about what plans each agency had in the area of space science for the following year. Summarizing an event that occurred presents an illustrative parallel to the current situation in China. When it came time for the Soviet representative to speak, to the shock of some others in the room (because of the known dire economic situation in the Soviet Union at that time) he confidently presented a spectacular and exciting array of approved plans. At the end, one of the Americans raised his hand and asked in wonderment and, likely, a hint of envy, if he had understood correctly, that the plans were really approved. The Soviet speaker confidently assured him that the plans were indeed approved. The American pressed further, asking then if the plans were approved and funded. The Soviet laughed heartily. The plans, he explained, were indeed approved, but funding was highly unlikely.

One gets much the same feeling about speeches made about space endeavors planned for the future in China. There may be a considerable gap between plans and reality. Present activities and focuses, as indicated by funds actually expended in support of an area, show that the Chinese are in a totally pragmatic mode, necessary as part of their constant quest to maintain internal stability. Returning to the theoretical models presented in Chapter 2, the evidence that the Chinese have pursued routes which keep them centered between capacity and equality is clear. They have little option but to follow rational decision making in pursuit of their goals. In that respect, the predictability of Chinese policy is likely higher than in many other countries. That being the case the challenge becomes identifying what trade-offs will be made concerning space in pursuance of their macro national goal of stability, and how the United States and other countries might influence both the trade-offs and the possible consequences. At the same time, there is also the 1.2 billion person potential market that is luring foreign business into opting for short-term advantage over long-term considerations to be considered. The further challenge then becomes to sort through the indicators and goals of all parties, short-term and long-term, to find points of overlap.

The Future

On 17 March 1996, the Fourth Plenary Session of the Eighth National People's Congress adopted the Outline of the Ninth Five Year Plan for National Economic and Social Development, and Long Term Targets for the Year 2010.[1] Space is considered within these endeavors. In formulating these targets, three specific challenges were seen as critical for consideration. First is that of increasing domestic demand for space-related services, telecommunications specifically, but also including remote sensing for land management and disaster warning and mitigation, and an autonomous satellite navigation system. Second, competition from the already established space powers, the United States and Russia in particular, is seen as a challenge. The third challenge foreseen by the Chinese is that in trying to reach the first goal, domestic markets can become controlled or occupied by foreign space companies. On the last point, the Chinese cite the figure that 80% of the Chinese domestic transponder market is occupied by foreign satellites and that currently, services and operations in the fields of meteorology, navigation, and remote sensing are all essentially dependent on foreign satellites. They fear that there is "a danger of losing whole domestic markets if China cannot build and launch its satellites better, faster and cheaper."[2]

With these challenges in mind, the Chinese then set out their policies for accelerating the pace of space development. These policies include points to:

- Support and expedite development of space technologies by the state.
- Enhance centralized and unified leadership and improve management.

Included in this section are statements regarding the desired integration of military and civilian applications, and to the need to "investigate and formulate a series of laws, regulations, systems and procedures to make the management work standardized and scientific."[3] Mention is also specifically made in support of competition for contracts within China.

Each point is interesting from many perspectives. For example, emphasis placed on an autonomous navigation system draws concern from some quarters because it is purported that several countries, including China, are actively engaged in exploiting the Global Positioning System (GPS), possibly for missile-guidance purposes. China has indeed shown interest in the integration of GPS into missiles and unmanned air vehicles and appears to be trying to acquire DPGS to improve the quality of their photogrammetric techniques. Here is where the autonomous navigation system fits in. Called Twin Start, deployment is planned for 1998 and it will have 20-meter accuracy.[4]

The point made about internal competition for contracts in China is also interesting. Although the Chinese there seem to acknowledge that the legacy of their planned economy and communist system is inefficiency and complacency, the example of the CAST and SAST attempt at competition for the Sinosat contracts shows that acknowledging a problem and being able to deal with it can be radically different. In more than one area, the albatross of the past seems to weigh heavily on China's ability to help itself.

Chinese priorities for the future in space will be pragmatic. In the civil area, it is highly likely that since commercial launches provide the funds to finance a variety of other space ventures it will be the number one priority. Reestablishing reliability must take immediate priority over new launcher development, with serialization of the LM then also remaining a near-term goal. Commercial satellites will likely also be supported and remote sensing capabilities expanded. Beyond this, everything becomes nebulous: manned space, reusable launch vehicle, planetary exploration, etc. In the military realm, the Second Artillery remains the favored child of the PLA, with emphasis likely to be placed on hardware with the potential for aid in military modernization efforts as well as for export sale. Cross utilization of technology will conceivably also not just continue, but increase. Indeed the DF-15, for example, may benefit in the future from integration of Global Positioning System guidance improvements.

Launchers

Until bookings for the Long March manifest are securely and consistently on the upswing it can be expected that the development of new vehicles will be significantly slowed. Work on their heavy lift vehicle, however, has apparently resumed. From descriptions, the technology is indicated to be similar to the sophisticated Ariane Vulcain and Japanese LE-5 engines.[5]

Militarily, strategic forces are the number one budget allocation within PLA. They are moving from a minimum deterrent strategy, primarily in response to the Russians, to one of limited deterrence with more flexibility, including tactical weapons and neutron bombs, now in response to both Russia and the United States. There are concerns that China is trying to upgrade its ballistic missile fleet with foreign technology, particularly with multiple independently targeted reentry vehicles (MIRVs).[6] That it has already successfully accomplished multipayload launches is certainly a step toward that capability. Talk in the United States on Theater Missile Defense (TMD) is very distressing to the Chinese, just as SDI was in the 1980's, as something they did not and do not have the technical ability to respond to, and which therefore has the possibility of rendering the PLA arsenal virtually obsolete.

Satellites

With telephone service available to under 5% of the Chinese population, telecommunications is and will remain a priority for Beijing. Plus, with the reversion of Hong Kong, China's investment stake in Intelsat, which is the world's largest provider of satellite-based telecommunications services, nearly doubled. Chinese interest now extends beyond Chinese borders from a business perspective. Government-supported aerospace concerns are under constant pressure to generate revenue as both evidence of their utility and hence justification for the government support, and as a supplement to the still decreasing amount of government support available. It is widely believed that CAST would like to build communications satellites for commercial sale. There are multiple problems, however, which will have to be first overcome.

First, the desire to sell abroad for currency generation will have to be balanced with the need to fulfill domestic communication needs. This relates to the Chinese policy goal of taking its own domestic satellite market back from foreign domination. All three goals—hard currency, domestic communications, and market independence—contribute to the stability equation which pervades all policy planning. It is hence likely that tangible government support for pursuing this objective will be forthcoming, with the understanding that the commercial sales aspect will take the longest to reach fruition because of the other obstacles to be overcome.

The other obstacles are technical. The Chinese are not currently economically efficient enough to produce for the international market. Also, the satellite technology is moving so quickly in some fields, like solar cells, that the Chinese cannot keep up. For example, although the Chinese have done some very good work with silicon in that area, most Western industries have already moved on to gallium arsenide. This means that the Chinese would still have to buy major components from abroad, as is the case with the DFH-3, to keep up with the international market. This quandry of "which satellite" also reflects back to the internal CASC-MPT disagreement over whether to utilize a DFH design (built by CAST, favored by CASC) or a new design relying on Western components (MPT favored). Since CASC is known to have championed CAST before, in the Sinosat "competition," it can be expected that it will continue to do so. Whereas MPT is looking for satellites offering the highest possible performance, CASC is looking to provide jobs and perpetuate its own bureaucratic position, as well as satellite performance.

This CASC-MPT disagreement about how to proceed in the future will be further complicated by the Global Mobile Personal Communications Satellites (GMPCS). Introduction of GMPCS services through such systems as Iridium, Globalstar, Odyssey, and others could be perceived by organizations such as the MPT as threatening its bureaucratic power base, and subsequently resisted.[7] Having China launch some of those satellites could mute some resistance from CASC, though whether that incentive would appease MPT is not as likely.

China has had some success at producing space batteries and has long experience in that area. In 1992, Sweden first used Chinese nickel-cadmium space batteries in their Freja satellite. Performance reports for the 6-ampere battery have been good and the Chinese have a second project with Sweden ready for launch (on a Russian launcher) in late 1997 or early 1998. Brazil also purchased two solar cells for the Brazilian satellite DCS-2 from the Chinese Space Power and Development Institute. In 1996, Lockheed-Martin asked to look at the Chinese batteries for possible purchase. The Chinese indeed sent twelve 80-ampere batteries to Lockheed-Martin in 1997 for testing, a considerable increase in capability from those used in Freja. With this kind of technical encouragement, a case could likely be made for further resources as there is an identifiable potential for both near-term and long-term payoffs.

Remote Sensing

Although conventional wisdom has it that civilian space programs have little military utility,[8] there are examples that seem to show otherwise, remote sensing being one of them. Critical parameters and performance figures for the major subsystems and components frequently overlap or are identical in civil and military space systems, such as imaging systems which are moving toward higher resolution and faster data delivery.[9] For example, the commercial products of SPOT and Landsat satellites were used extensively in Operations Desert Shield and Desert Storm for broad-area search and mission planning. Geographic Information Systems (GIS), comprising personal-computer hardware and very sophisticated software (e.g., AutoCad), now permit users to make very accurate digital maps with GPS data outputs. One can use such hardware and software capabilities for more than just preprogramming the route of a cruise missile. Better maps and commercially available satellite imagery allow Third World states to develop better targeting by improved photogrammetric techniques.[10]

China's plans are clearly ambitious. Their challenge is to match ambitions with resources and technology. Not surprisingly or nefariously, that will likely be their primary motivation for entering into any international cooperative project.

Analysis and Recommendations

In Chapter 1, I stated that I would later extrapolate from the information provided and project about China's future and equally important, what policy actions the United States might take to avoid a confrontational stance with China while encouraging Beijing to build a more stable, cooperative regime. The importance of avoiding a confrontational stance stems from attitudes and actions in both China and the United States. As a developing state, China is experiencing growing pains which can result in either maturation or anxiety and subsequently defensiveness and hostility. Obviously, it would behoove the United States to encourage the former. Perhaps that Chinese President Jiang Zemin during his October-November 1997 trip to the United States admitted, though ambiguously, that "mistakes" had been made in dealing with internal problems[11] could be viewed as a promising sign in that regard. However, a poll conducted during that same period found that 36% of Americans asked consider China unfriendly or an outright enemy, while only 25% consider it a friend or ally.[12] Therefore, for at least some Americans, Chinese actions will have to more closely match their rhetoric before an opinion shift occurs from the "confrontation-adversarial" side of the political spectrum to the more "cooperative-friendly" end, with the latter more conducive to international stability. Clearly, if that diversity of opinion exists in the United States, it is likely also found in China. Confidence-building measures are therefore desirable and utilitarian for both.

The Chinese consider that they have "approached and reached the advanced world level" in areas of technology including: satellite recovery; the use of cryogenic propellants; geostationary satellite launches; measurement and control; multipayload launches; and strap-on boosters.[13] They have a well-deserved confidence and pride in the their space program, the recent launch failures notwithstanding. Because of their technical abilities in the referenced areas and their commercial launches, they are often included among the ranks of the space programs of the United States, Europe, Russia, and Japan. This placement allows the Chinese the status and respectability that they feel they deserve and that they badly want. On the other hand, in order to sustain that program, advance it to a technical level where they can be internationally competitive in areas like commercial satellites, and keep their program moving forward, they are under considerable pressure not only to generate commercial launch sales, but also to supplement their funds and knowledge from the outside.

Because of the market potential China has to offer, there seems little doubt that foreign businesses will be willing to make considerable concessions to China in terms of both technology transfer and modes of acceptable business operations in return for a foothold into the market. Businesses engage in risk management, rather than the risk aversion which seems to have overtaken the government of late.[14] They clearly have decided that the potential near-term benefits of the Chinese market more than offset the risks emanating from the *Guan Xi* laden environment. U.S. businesses are more restrained by government regulations than others, but all will be posturing for position.

With China so clearly concerned about foreign domination it seems highly likely that it will encourage foreign businesses into China as long as they are useful and needed, but not to an extent where they would in either the near term or long term become an influence on Chinese policy. Although it would be desirable from a governmental perspective and prudent from a business perspective to hope that foreign business would try to gradually impose their expectations concerning legal and business practice standards on the Chinese, it is unlikely. Near-term profit potential and the tendency to rely heavily on Chinese nationals as company representatives, individuals not only more comfortable with a *Guan Xi* approach to doing business but aware that it benefits them as well, will likely further perpetuate the system. Therefore, it seems up to governments to persuade Beijing that in order to sustain a long-term relationship with the outside world, it is in its best interest to make certain changes. Suggestions on how this might be done fall into three general categories: implementation of a legal system, demanding more reciprocity, and international cooperation. They are offered as "confidence-builders" for both China and the United States. Ideally these would be carried out sequentially, in the order given. As that is not likely possible, however, they are discussed more in order of near-term to longer-term potential policy courses.

Reciprocity

Looking at the past, it seems useful to examine at least one instance where persuasion techniques were successful in moving China to acceptance of a position or undertaking an action it might not have otherwise. Getting China to agree to adhere to the MTCR guidelines seems to provide such an example. Speaking to the press in Washington the Chinese Ambassador to the United States Zhu Qizhen made some interesting comments about an arms sale to Pakistan. "We have sold some conventional weapons to Pakistan, including a . . . amount of short-range tactical missiles. I think you here call it M-11. We don't call it M-11, but since you say M-11, let's say M-11 . . . We don't use the name M-11. It is a United States code name, M-11. Let's say if it is M-11 this is within the range of the MTCR; that is the range is only a little more than 200 kilometers."[15] The comments are interesting, and somewhat ironic, as illustrative of how Chinese secrecy can at times even complicate things for the Chinese. Indeed M-11 was the Chinese, not the American, code name for the missile. But, as he points out, it could be that Pakistan bought a missile other than the M-11! What he was really trying to point out in this convoluted monologue was that China had not violated the MTCR guidelines. He felt compelled to make this point because if a violation was found to have occurred sanctions could have followed.

China was persuaded to abide by MTCR restrictions in February 1992. This concession was made under considerable duress, and some feel at the cost of a deterioration in Sino-American relations. It was made following

Washington's promise to lift sanctions on the sale of satellite parts and high-speed computers to China. Syria was to be the first customer for the Chinese M-9 missile, but that sale would have violated the MTCR agreement to restrict sales beyond the 300 km/500 kg capability. The Chinese agreed to stop the sale, as other commodities they wanted were at risk because of the M-9 and apparently were more important. Giving them something that they want for something that we want was effective: reciprocity. This is a concept well recognized by most politicians worldwide, but one with which Beijing is not always comfortable. It is usually dismissed under the guise of either claiming China is a developing country and therefore unable to respond in the same way as other developed countries, or that China is a developed country and indeed a world power, invoking sovereignty to simply refuse. This is part of the schizophrenic nature of Chinese foreign policy.

As a further example, although Chinese officials, military and nonmilitary, regularly come to the United States and are given guided tours of nearly every facility requested by them (assuming the facility is available to foreign delegations generally), the Chinese are not nearly so amenable to U.S. delegations visiting China. It is only after significant prodding, successful only some of the time, that officials are taken beyond the Great Wall and the terra-cotta soldiers, to space centers and air bases. Because of the benefits to potentially be yielded in terms of more accurate assessments of Chinese capabilities, pressing more frequently and harder for reciprocity from Beijing deserves further attention.

In general, a policy of reciprocity at the macro level may be a viable way of allowing Beijing to achieve its own internal goals while at the same time tolerating U.S. pursuance of its own priorities. That would, of course, require that the United States set some policy priorities rather than simply responding to the demands of whatever interest group got the most attention in the press the day before a decision is made. Assuming that hurdle could be overcome on at least some issues, it would seem that the next step would be to inform Beijing of the policy, the rationale, and the issues that the United States is least willing to bend on. This is important for "face" purposes: allowing to plan where and how they can make concessions gracefully, or not so gracefully. Indeed the United States ought not be surprised by some harsh public rhetoric from Beijing, similar to the old Soviet style. Obviously, the political downside of such an approach in the United States would be taking the heat from supporters of those issues which do not make the priorities list.

Reflecting on the Mabuhay Philippine satellite pricing disagreement discussed in Chapter 5, clearly the Clinton administration has been willing to endure criticism from the hard-line China politicians and press in order to gently push them into alignment with other nations, rather than risk perhaps shoving them into a more defensive mode than necessary and likely losing whatever movement toward international standards has been gained. Realistically, this approach seems to maximize the influence leverage available. Rather than merely utilizing this approach as opportunities arise, however, an active effort ought to be made to generate opportunities. This might be done through international cooperation, something the Chinese have repeatedly stated their interest in increasing, not only for the very practical benefits to be yielded, but also for the political prestige as an equal as well.

International Cooperation

There is considerable evidence that because of the gap China still suffers regarding space, in terms of advanced technology and systems engineering, it is still primarily responsive to policy and strategy rather than innovative. This responsiveness dates back to its early space program. While technologically the Chinese designers had to follow the Soviet path of building ballistic missiles with liquid propellants and large throw weight, they learned of the main strategic trends almost exclusively from the West because Soviet writings rarely revealed valid details on Soviet nuclear strategy. When, for example, the Western media reported that the United States had been developing an antiballistic missile system mainly for protection "against the small ICBM force that the Chinese might develop in the future," Beijing's designers urgently concentrated on the penetration capability ("penetrability") of their ICBM in addition to its range and accuracy.[16] This strategic responsiveness, coupled with the technological need to sometimes model their facilities and hardware after that already proven (e.g., siting their launch facilities at the same latitude as Kennedy Space Center, mentioned earlier) and their stated desire for international cooperation, seems to offer opportunities for influencing Chinese actions in subtle ways.

Considering that Beijing has shown a propensity to respond to actions and strategies rather than initiate them, an approach which their current economic situation seems to support now as well, and that they would like to move a significant portion of their workforce into commercial areas, one might then look at the history of space cooperation and take a lesson from the past. When the space programs of the United States and the Soviet Union were locked in political competition that translated into a space race, many European countries saw the need to become involved or get left behind. Indeed there is considerable European "technology gap" literature from that period discussing the dilemma.[17] Canada had similar concerns. When it was clear that other countries and groups of countries intended to get involved in space activity, the prudent policy choice for the United States was to engage them in cooperative activities so as to guide the directions they took.[18] This was not nefarious on the part of the United States; it simply created a win-win situation for both. Other countries got experience in space much sooner than they would have otherwise and the United States was able to have significant influence, initially, concerning where and how. Perhaps the United States ought to take that general approach again.

Traditionally, the one area of space cooperation which has been the least threatening from a national security perspective and the most fruitful in terms of goodwill benefits has been space science. Scientists are driven by goals defined by nature rather than governments. They also tend to share the results of their findings through publications as quickly as possible anyway. That means that although basic science can progress faster or slower in one country than another, it cannot be patented or kept from others. Physics is physics, regardless of whether experiments are performed in Cambridge, Massachusetts, or Cambridge, England, Beijing or Moscow. Basic science is also an area stated by Song Jian as important to China, but not currently considered a pragmatic area for immediate pursuit. Perhaps the United States ought to make directed initiatives to Beijing in the areas of space science research, to be performed in China. This would serve the purpose of creating goodwill for the United States, guiding Chinese expenditures and efforts into this area whereas otherwise that might not be the case, and likely generate some very interesting and useful science data. In some ways, it is analogous to the sums of money being poured in Russia today through the ISS program, as a more utilitarian answer to helping the Russian economy than direct foreign aid. The difficulty might well come in that the space science budget must constantly fight to hold its ground against the developmental and operational costs of ISS. So, rather than cut further from the livelihoods of American space scientists, cooperative programs in space science with China ought to be created as a supplemental program or funded through prioritizing work with China through federal grants, not from the regular NASA space science program.

To take this idea of using cooperative programs to influence the direction of Chinese policy, the case of the United States allowing the transfer of launch technology to Japan beginning in 1969, though it had denied the same to Europe and Canada and it was against all policies, might be also considered.[19] After much consternation from both NASA and DOD, McDonnell-Douglas was allowed to license Delta launch technology as part of a political effort to strengthen political ties between the United States and Japan. Although at the time there was considerable concern that Japan would immediately become a launch competitor to the United States that did not happen for many years, if one considers that it has happened now through the H-2 launcher. Indeed the largest competitor to the U.S. launch vehicles, the Ariane, was created after and at least partly because the United States wanted to control what the Europeans could launch on U.S. boosters, specifically communication satellites.[20] Here, competition was generated due to a restrictive U.S. policy rather than a cooperative one.

Perhaps the United States ought to allow the licensing of *selected* technology in China, perhaps for producing all or part of domestic satellites or in some other field deemed appropriate.[21] If satellites were produced or jointly produced in China under a license agreement, they could not be sold for export without U.S. approval, reducing the impact on U.S. businesses. Clearly, there would need to be careful consideration of what, to what extent, for how long, and the terms in general. In return for the technology, there would be guaranteed short-term returns to U.S. businesses for the satellites they might otherwise build in their entirety, there would be jobs involved for the Chinese, and perhaps most importantly, there would be the opportunity to guide their program development. Although loss of U.S. jobs is certainly a negative consideration, at least theoretically the money returned to the U.S. companies could be reinvested toward creation of new products to create new jobs.

Obviously, one of the big concerns of the United States is that technology transferred for one purpose will be put to another use. Indeed in April 1997, machine equipment purchased from McDonnell-Douglas Corporation in 1994 for civilian aviation purposes was found to have somehow wound up at a military complex in Nanchang which builds missiles and fighter aircraft. Catic, the Chinese-government-owned company which bought the equipment, was purported to have deliberately misled Clinton administration officials which had approved the sale. Charges that trade interests were put above national security interests were asserted against the administration,[22] which will likely increase hesitation about these types of arrangements. Nevertheless, controlled licensing of technology in conjunction with the next recommendation seems to offer the United States more leverage than a free-wheeling technology grab by the Chinese from global companies perhaps overly anxious to establish their place in the Chinese market.

Legal Aid

Finally, we return to the issue of culture. The Chinese culture is to them, admittedly and proudly, Sinocentric. They are the Middle Kingdom and the rest of the world hovers on the periphery. Although they have been down and out lately, they have no doubt that things will change and go back to the way of the past (Confucianism) and that China will again be the Center of the Universe. If that is to be believed, and with the longest continual history on Earth the Chinese have no reason not to believe it, then the Chinese are not really looking to integrate themselves with the Family of Nations. Instead, it might be posited that they are seeking to learn how to employ the tools and resources of the Family of Nations for the benefits they yield, in order to restore themselves to their rightful place. This by no means implies miscreant intent, but merely points out that there is an important long-term difference between integration and adaptation, with the Chinese traditionally leaning toward the latter not the former. If that is the case, then it would greatly benefit other countries to integrate China into the mainstream whenever possible, so that it might share not only in the benefits of the tools and resources that development brings, but the responsibilities as well.

The Chinese bemoan constantly how much they would like to have a developed, mature legal system to accommodate business transactions as the rest of the world does but, they explain, they are a developing country. As such a legal system would significantly level the commercial playing field in China and would hence be of great benefit to U.S. companies, aerospace and otherwise, it seems not only logical but imperative that the United States flood the Chinese with all the assistance they could need in setting up such a system, and the sooner the better. As the largest of the emerging markets, China's greatest clout now and likely in the near term will be through economics. Continued use of the *Guan Xi* system excludes or at least minimizes the influence that foreign business concerns would otherwise have in a situation such as this, and it is a system which will not change without significant outside influence. The Chinese claim that they want to move from a primarily *Guan Xi* driven system to one more in line with the international business community. It would likely be akin to the Japanese system where legal parameters temper the impact of *Guan Xi*. The United States will be seriously negligent if it does not encourage this move in every way possible. Imparting the tools and knowledge to institute a legal system is imperative. As instituting a viable legal system means changing Chinese culture it will likely be difficult if desired, and nearly impossible if resisted, tacitly or implicitly. Also, the fact that instituting even commercial legal parameters into China could have spillover effects on the authoritarian government is likely not lost on the government. Nevertheless, until and unless Beijing retracts its stated intent of improving its legal and contractual framework for business operations, assisting it toward that end is a prudent course for the United States to pursue.

Even acknowledging the cultural propensity against such, laws and the legal system did not escape the wrath of the Cultural Revolution and its targeting of anything "establishment." Legal education virtually stopped during that period.[23] Courts became political tools, and legal training was not necessary to implement "the people's law," only a sense of political correctness. Now, however, a modern legal system is badly needed to accommodate economic reform, particularly foreign investment. With the Hong Kong stock market plunging 10.6% on 23 October 1997[24] and analysts evaluating the impact the drain of capital from new Chinese enterprises to old Chinese problems played in that debacle, it becomes even more imperative.

The Ford Foundation has several "legal aid" programs currently underway in China, including: two grants to help reform the Chinese criminal justice system; one grant to teach new methods and material in criminal procedure law and judicial implementation of administrative law; sponsorship of a conference on legal reform; and the development of two books on proposed legislation to use the law in China to regularize government procedures. Though admirable, much more could be done. The Ford Foundation grants to China for the above mentioned programs total under $300,000, and focus primarily on criminal law because of the human rights violations incurred in conjunction with questionable criminal law practices.

Establishing a civil law code ought to be addressed also, assisted by the United States on a significant scale to include bringing Chinese lawyers to the United States for training and sending legal teams to China to assist with developing the legal codes that the Chinese claim they want to bring them closer into the Family of Nations. From civil law, expansion of written rules might spill over into criminal law as well. At least the expertise to begin to think about it would be more established. From a U.S. perspective, such a program seems to have few downsides: the United States would be giving China something it claims to want and it would benefit the United States as well. From a Chinese perspective, those who are familiar with Western culture may see it as the West cleverly releasing a plague of locusts wearing business suits on them, but they have asked for this particular plague.

The same type of tutorial program in areas like establishment of a free press would also be desirable. Although the press is not officially censored, access to information is tightly controlled and limited by unseen and unwritten, but still powerful, boundaries. For example, when Deng Xiaoping died some 10,000 people gathered to mourn him at a statue of Mao in the center of Chengdu. The next day, Chinese police lined the streets to "discourage" people from any further activity. Neither event was ever noted in the newspaper, in Chengdu or elsewhere. If the press does not report it, it simply did not happen. Moving away from this Orwellian view of information handling can perhaps be achieved best and fastest beginning by dialogue with those with some experience. There may be other areas where the United States might assist China too, rather than the United States simply demanding implementation of a system based on the U.S. model.

Slowly but consistently, China ought to be nudged toward building commonalities with other countries. In many ways, it is still uncomfortable with the idea of coming out of the isolation it has always preferred. But integrating China into sharing the responsibilities of a great power, as it so desires to be again, as well as sharing the benefits, ought to be continually stressed. Moving from confrontation to cooperation is a process of integrating, not just adapting.

Conclusions

The thesis in Chapter 1 stated that because China is at a precarious point of development, careful analysis is warranted within the U.S. policy community toward identifying ways in which the United States might urge China toward a more acceptable position on the international political spectrum. Specifically regarding space, China has traditionally emphasized the military and arms sales potential of space and space hardware. The United States ought to focus on the development of constructive U.S. policies aimed at encouraging China to alternatively emphasize space as an element of peaceful domestic development. These polities ought then be vigorously pursued by the United States. The time is now propitious because China's current international stature comes more from its market size than its military position or ambitions, perhaps making it more receptive to outside influence than might be the case in 10 to 15 years if the status quo prevails.

I have attempted to present a beginning for that process. Although at times it may have seemed that the book was as much or more a sociopolitical analysis of China generally, rather than about space, I would suggest that such a broad analysis is necessary and inherent because of the nature of the Chinese culture and the strategic importance space holds within China politically and economically.

Many Chinese "space leaders" were opposed to the violent crackdown in Tiananmen Square and appalled by it. They were, however, equally appalled by what many saw as the West's arrogant and hypocritical response to it. Like others, they want self-respect. "Space" is but one piece of the puzzle which makes up China and to consider it individually would result in missing the influences of and implications on the bigger picture.

China has changed dramatically in past years. It is truly one of the most dynamic countries in the world. In some ways, however, the environment within which space research, development and even operations take place has changed little and is still constrained to an almost untenable degree. The bureaucracy is stifling, resources meek, and personal freedoms narrowly defined. One would need a more confident, prescient sense than is judicious or typical in China today to step beyond the immediate dictates of getting through the day by maintaining and supporting the status quo, and move toward innovativeness. Innovation is part of the Chinese puzzle which is still missing. What impact that human factor has on the ability to extrapolate about future plans in China is perhaps the hardest factor to include. Indeed an analyst would truly have to be an oracle to accomplish such, which is certainly not the case here. The almost surrealistic nature of China today, imposing an ancient culture on a modernizing society overshadowed by remnants of a gnarled recent past, assures that it will be a dynamic force internationally for many years to come, and certainly one requiring continual monitoring and nurturing into the Family of Nations.

Endnotes

1. "Fast-track development of space technology in China," *Space Policy*, May 1996, 139.
2. "Fast-track development of space technology in China," *Space Policy*, May 1996, 139.
3. "Fast-track development of space technology in China," *Space Policy*, May 1996, 140.

4. Dennis M. Gormley and K. Scott McMahon, "Proliferation of Land-Attack Cruise Missiles: Prospects and Policy Implications," in Sokolski, 143.
5. Craig Covault, "Chinese Manned Flight Set for 1999 Liftoff," *Aviation Week & Space Technology*, 21 October 1996, 22.
6. Joseph C. Anselmo, "U.S. Eyes China Missile Threat," *Aviation Week & Space Technology*, 21 October 1996, 23.
7. Steve Watkins, "Asian Nations Resist Opening Doors to Global Phone Services," *Space News*, 15–21 September 1997, 4,32.
8. Dennis M. Gormley and K. Scott McMahon, "Proliferation of Land-Attack Cruise Missiles: Prospects and Policy Implications," in Sokolski, 144.
9. Berner, 95.
10. Gormley and McMahon, 144.
11. Seth Faison, "Beijing's New Face," *The New York Times*, 3 November 1997, A1.
12. Barbara Slavin, "Poll: Many Consider China an Adversary," *USA Today*, 29 October 1997, 5A.
13. Zhang Xinzhai, 1996, 24.
14. Joan Johnson-Freese and Roger Handberg, *Space, the Dormant Frontier: Changing the Space Paradigm for the 21st Century*, (Westport, CT: Praeger Books, 1997), 165–170.
15. Lewis and Hua, 37.
16. The quotation, from Robert McNamara, is found in Coit D. Blacker and Gloria Duffy, eds., *International Arms Control: Issues and Agreements*, 2nd ed. (Stanford, CA: Stanford University Press, 1984) 206, and cited by Lewis and Hua, 1992, 21.
17. Jean-Jacques Servan-Schreiber, *The American Challenge*, (New York: Avon Books, 1967); Roger Williams, *European Technology: The Politics of Collaboration*, (London: Croom Helm, Ltd.,1973).
18. Joan Johnson-Freese, *Changing Patterns of International Cooperation in Space*, (Malabar, FL: Orbit Book Co., 1990), Chapter 1.
19. Joan Johnson-Freese, *Over the Pacific: Japanese Space Policy into the 21st Century*, (Dubuque, Iowa: Kendall-Hunt Publishing, 1993), 125–126.
20. Joan Johnson-Freese, 1990, 25–26.
21. As of spring 1998, apparently the Clinton administration is considering such a policy.
22. Gary Milhollin, "China Cheats (What a Surprise!)," *The New York Times*, 24 April 1997, A29.
23. Hungdah Chiu, "Institutionalizing a New Legal System in Deng's China," in *China in the Era of Deng Xiaoping: A Decade of Reform*, Michael Ying-Mao Kau and Susan H. Marsh, eds., (Armonk, NY: M. E. Sharp, 1993) 62.
24. Sara Webb, "Asia's Storm Isn't Over, Many Say," *The Wall Street Journal*, 27 October 1997, C1.

Glossary

AVIC	Aviation Industry of China
CAAC	Civil Aviation Association of China
CAC	China Aerospace Corporation
CALT	China Academy of Launch Vehicle Technology
CASC	China Aerospace Corporation
CAST	China Academy of Space Technology
CBERS	China-Brazil Earth Resources Satellite
CCTV	China Central Television, Channel 1
CETV	China Education Television
CGWIC	China Great Wall Industry Corporation
CITIC	China International Trust and Investment Corporation
CLTC	China Satellite Launch and Tracking Control
CNSA	China National Space Administration
COMSAT	Communication Satellite Corporation
COSTIND	Commission of Science, Technology and Industry for National Defense
CPMIEC	China Precision Import-Export Company
CSS	Chinese surface-to-surface missile
CZ	Chang Zheng (Chinese for Long March)
DBS	direct broadcasting system
DF	Dong Feng (East wind) program of Chinese land-based missile development
DFH	Dong Fang Hong (The East is red). Chinese geostationary satellites originally referred to by the acronym STW
DOD	(U.S.) Department of Defense
DOS	(U.S.) Department of State
ERS	Earth Resource Satellite
ESA	European Space Agency
FB	Feng Bao (Storm) missile
FSW	Fanhui Shi Weixing (return test) satellite. Also called Jianbing or Progress

FY	Feng Yun satellite
GEO	geostationary Earth orbit
GIS	geographic information systems
GMPCS	Global Mobile Personal Communications Satellite
GNP	gross national product
GPD	the arm of the Communist Party within the PLA
GPS	Global Positioning System
GTO	geostationary transfer orbit
Guan Xi	A complex web of relationships by which interactions are defined (pronounced "Gwan-Shee")
IAF	International Astronautical Federation
ICBM	intercontinental ballistic missile
IRBM	intermediate range ballistic missile
ISS	International Space Station
ITAR	International Traffic in Arms Regulations
JSLC	Jiuquan Space Launch Center
KCS	Kennedy Space Center
LEO	low Earth orbit
LM	Long March (launchers)
MAI	Moscow Aviation Institute
MASI	(Chinese) Ministry of Space Industry
MC	munitions control
MFN	most favored nation trade status
MFRT	Ministry of Film, Radio and Television
MIRV	multiple independently targeted reentry vehicle
MPT	(Chinese) Ministry of Post and Telecommunication
MTCR	Missile Technology Control Regime
NASA	(U.S.) National Aeronautics and Space Administration
NDSTIC	(Chinese) National Defense Science, Technology and Industry Commission
NRO	(U.S.) National Reconnaissance Office
NRSC	(Chinese) National Remote Sensing Center
ODTC	(U.S.) Office of Defense Trade Controls
P&T	post and telecommunications
PICC	People's Insurance Company of China
PLA	(Chinese) People's Liberation Army
PRC	People's Republic of China
R&D	research and development
SAST	Shanghai Academy of Spaceflight Technology
SD	smart dispenser
SJ	satellites for space physical exploration (Shi Jian, or experimental)
SLG	(Chinese) Space Leading Group in the State Council
SOE	(Chinese) state-owned enterprises

SSO Sun-synchronous orbit
STW Shiyan (experimental) Tongxun (communications) Weixing (satellite)
TMD theater missile defense
TT&C tracking, telemetry, and control
Wei Qi a strategic game played in China, somewhat like chess (pronounced "Way-Chee")
WTO World Trade Organization
Xitong literally means "system" (of organization)
XSCC Xi'an Space Command and Control Center
XSLC Xichang Space Launch Center

Selected Bibliography

Books

The APT Yearbook, 1997, A. Narayan, ed., (Surrey, UK: ICOM Publications Ltd., 1997).

Chang, Iris. *Thread of a Silkworm*, (New York: Basic Books, 1995).

China Today: Defense Science and Technology, Vol 1., (Beijing: National Defence Industry Press, 1993).

Clark, Phillip, ed. *Jane's Space Directory, Thirteenth Edition*, (Coulsdon, UK: Jane's Information Group Ltd., Sentinel House, 1997).

Elvin, Mark. *The Pattern of the China Past*, (Stanford, CA: Stanford University Press, 1973).

Garside, Roger. *Coming Alive: China After Mao*, (New York: McGraw-Hill Book Company, 1981).

Goodrich, L. Carrington. *A Short History of the Chinese People*, (New York: Harper & Row, 1943).

Handberg, Roger, and Joan Johnson-Freese. *The Prestige Trap*, (Dubuque, IA: Kendall-Hunt Publishing, 1994).

Haw, Stephen G. *A Traveler's History of China*, (New York: Interlink Books, 1995).

Johnson, Nicholas L., and David M. Rodvold. *Europe and Asia in Space, 1993–94*, Prepared for USAF Phillips Laboratory, Kirtland AFB, NM, by Kaman Sciences Corporation, Colorado Springs, CO.

Johnson-Freese, Joan. *Changing Patterns of International Cooperation in Space*, (Malabar, FL: Orbit Book Co., 1990).

Johnson-Freese, Joan. "*Over the Pacific: Japanese Space Policy Into the 21st Century*," (Dubuque, IA: Kendall-Hunt, 1993).

Johnson-Freese, Joan, and Roger Handberg. *Space, the Dormant Frontier: Changing the Space Paradigm for the 21st Century*, (Westport, CT: Praeger Books, 1997).

Lewis, John Wilson, and Xue Litai. *China Builds the Bomb*, (Stanford, CA: Stanford University Press, 1988).

Lieberthal, Kenneth. *Governing China*, (New York: W. W. Norton & Company, 1995).

McDougall, Walter. "*. . . the Heavens and the Earth*," (New York: Basic Books, 1985).

Meisner, Maurice. *The Deng Xiaoping Era*, (New York: Hill and Wang, 1996).

Peyrefitte, Alain. *The Chinese, Portrait of a People*, translated from the French by Graham Webb, (New York: Bobbs-Merrill Company, Inc., 1973).

Spence, Jonathan. *To Change China*, (Boston: Little, Brown, 1969).

Sokolski, Henry, ed. *Fighting Proliferation: New Concerns for the Nineties*, (Maxwell AFB, AL: Air University Press, 1996).

Tuchman, Barbara. *Notes From China*, (New York: Collier Books, 1972).

Articles

"Agila 2 Uses Extra Fuel to Reach Proper Orbit," *Space News*, 8–14 September 1997, 2.

Anselmo, Joseph C. "U.S. Eyes China Missile Threat," *Aviation Week & Space Technology*, 21 October 1996, 23.

"AsiaSat Seeks $58 Million on Insurance Claim," *Space News*, 16–22 September 1996, 3.

Buruma, Ian. "Taiwan's New Nationalists," *Foreign Affairs*, July/August 1996.

Blustein, Paul. "The China Syndrome: Despite the Criticism, engaging Beijing may be the United States' only option," *Washington Post National Weekly Edition*, 21 April 1997, 23.

Bogert, C. "Pray for China," *Newsweek*, 9 June 1997, 44–45.

Chen, Yanping. "China's space policy, a historical review," *Space Policy*, May 1991, 116–128.

Chen, Yanping. "China's space commercialization effort, organization, policy and strategies," *Space Policy*, February 1993, 45–53.

Chen, Yanping. "China's Space Interests and Missile Technology Controls," *Space Power Interest*, Peter Hayes, ed., (Boulder, CO: Westview Press, 1996) 71–84.

"China Defeats a U.N. Resolution Criticizing Its Human Rights Record," *The New York Times*, 16 April 1997, A11.

"China's Growing Inequality: As Economy Takes Off, Millions Are Left Behind," *The Washington Post*, 1 January 1997, A1.

"China's 'Last Emperor,' " *The Washington Post*, 20 February 1997, A26.

"ChinaSat Payoff Begun," *Aviation Week & Space Technology*, 14 October 1996.

"Chinese Detail Small-Satellite Efforts," *Aviation Week & Space Technology*, 14 October 1996, 33.

Chiu, Hungdah. "Institutionalizing a New Legal System in Deng's China," in *China in the Era of Deng Xiaoping: A Decade of Reform*, Michael Ying-Mao Kau and Susan H. Marsh, eds., (Armonk, NY: M. E. Sharp, 1993).

Covault, Craig. "Chinese Manned Flight Set for 1999 Liftoff," *Aviation Week & Space Technology*, 21 October 1996, 22.

Covault, Craig. "China Seeks Cooperation, Airs New Space Strategy," *Aviation Week & Space Technology*, 14 October 1996, 29.

DeSelding, Peter B. "Chinese Set Ambitious Plans," *Space News*, 14–20 October 1996, 19.

Dornhein, Michael A. "DF-15 Sophisticated, Hard to Intercept," *Aviation Week & Space Technology*, 18 March 1997, 23.

Faison, Seth. "Beijing's New Face," *The New York Times*, 3 November 1997, A1.

"Fast-track development of space technology in China," *Space Policy*, May 1996, 139.

Ferster, Warren. "U.S. Says China Violated Accord," *Space News*, 19–25 May 1997, 18.

Filho, José Monserrat. "Brazilian-Chinese space cooperation: an analysis," *Space Policy*, May 1997, 153–170.

Fisher, Arthur. "A Long Haul for Chinese Science," *Popular Science*, August 1996.

Fluendy, Simon. "Up in Smoke," *Far Eastern Economic Review*, 5 September 1997.

Gargan, Edward A. "Hong Kong Erupts At Plan to Curtail Rights," *The New York Times*, 11 April 1997, A8.

Greenberger, Robert S. "Favored-Nation Status for China Loses Its Certainty," *The Wall Street Journal*, 15 April 1997, A20.

Greenberger, Robert, and David Rogers, "China Vote Divides GOP's Christian and Business Wings," *The Wall Street Journal*, 24 June 1997, A24.

He Changchui, "The development of remote sensing in China," *Space Policy*, February 1989, 65–74.

Kahn, Joseph. "China Has No Need to Suppress the Press in Hong Kong Now," *The Wall Street Journal*, 21 April 1997, A1.

Kaplan, Morton A. "Systems Theory," in *Contemporary Political Analysis*, James C. Charlesworth, ed., (New York: Free Press, 1967).

Karmel, Solomon M. "The Chinese Military's Hunt for Profits," *Foreign Policy*, Summer 1997, 102–112.

Lachica, Eduardo. "Chinese Official Says WTO Pace May Take Time," *The Wall Street Journal*, 18 April 1997.

Langereux, Pierre, and Christian Lardier, "Launch Setbacks fail to dent China's space ambitions," *Interavia*, December 1996.

Leiken, Robert S. "Controlling the Global Corruption Epidemic," *Foreign Policy*, Winter 1996–96, 55–76.

Lewis, John Wilson, and Hua Di. "China's Ballistic Missile Programs, *International Security*, Fall 1992, Vol. 17, No. 2) 5–40.

Lewis, John W., Hua Di, and Xue Litai. "Beijing's Defense Establishment: Solving the Arms Export Enigma," *International Security*, Spring 1991, 97–109.

Liu Ji-yuan, and Min Gui-rong. "The progress of astronautics in China," *Space Policy*, May 1987, 141–147.

Mecham, Michael. "China Plans Seven Missions For Long March Booster in 1997," *Aviation Week & Space Technology*, 11 November 1996, 25.

Mecham, Michael. "China Displays Export Air Defense Missile," *Aviation Week & Space Technology*, 2 December 1996, 61.

Milhollin, Gary. "China Cheats (What a Surprise!)," *New York Times*, 24 April 1997, A29.

Mufson, Steven. "Chinese Shake-Up Leadership, *The Washington Post*, 19 September 1997.

Pirard, Theo. "Space Business Around the World," *Spaceflight*, September 1997, 294.

Pike, Gordon. "Chinese launch services: A user's guide," *Space Policy*, May 1991, 103–115.

Riggs, Fred W. "The Theory of Political Development," *Contemporary Political Analysis*, James C. Charlesworth, ed., (New York: Free Press, 1967).

Schweitzer, Peter. "You, Too, May Be Funding China's Army," *USA Today*, 14 May 1997, A13.

Slavin, Barbara. "Poll: Many Consider China an Adversary," *USA Today*, 29 October 1997, 5A.

Tacey, Elisabeth. "Chinese Rocket Site 'Blind to Safety,' " *Nature*, 13 June 1996.

Taylor, Jeffrey. "China Puts Heat on U.S. Firms to Lobby for 'Most-Favored-Nation' Trade Status," *The Wall Street Journal*, 24 June 1997, A24.

Ward, Mark. "Exploding China's Dreams," *Interavia*, 16 March 1997.

Webb, Sara. "Asia's Storm Isn't Over, Many Say," *The Wall Street Journal*, 27 October 1997, C1.

Williamson, Mark. "Chinese Space Show," *SPACE and Communications*, November-December 1996.

Wu Bian, "Space Industry Promotes Modernization," *Beijing Review*, 6-12 January 1997, 14.

Wu Guoxiang, "China's space communication goals," *Space Policy*, February 1988, 41-45.

Wu Guoxiang, "China's space communication goals," *Space Policy*, February 1988, 42.

Yeung, Chris, et al., "Mourners Struggle to the End," *South China Morning Post*, 5 June 1997, 1.

Zhang Xinzhai, "The Achievements and the Future of the Development of China's Space Technology," *Aerospace China*, Summer 1996, 25.

Zhu Yilin, and Xu Fuxiang, "Status and prospects of China's space programme," *Space Policy*, February 1997, 69-75.

Index